The Aesthetics of Atmospheres

Interest in sensory atmospheres and architectural and urban ambiances has been growing for over 30 years. A key figure in this field is acclaimed German philosopher Gernot Böhme whose influential conception of what atmospheres are and how they function has been only partially available to the English-speaking public. This translation of key essays along with an original introduction charts the development of Gernot Böhme's philosophy of atmospheres and how it can be applied in various contexts such as scenography, commodity aesthetics, advertising, architecture, design, and art.

The phenomenological analysis of atmospheres has proved very fruitful and its most important, and successful, application has been within aesthetics. The material background of this success may be seen in the ubiquitous aestheticization of our lifeworld, or from another perspective, of the staging of everything, every event and performance. The theory of atmospheres becoming an aesthetic theory thus reveals the theatrical, not to say manipulative, character of politics, commerce, of the event-society. But, taken as a positive theory of certain phenomena, it offers new perspectives on architecture, design, and art. It made the spatial and the experience of space and places a central subject and hence rehabilitated the ephemeral in the arts. Taking its numerous impacts in many fields together, it initiated a new humanism: the individual as a living person and his or her perspective are taken seriously, and this fosters the ongoing democratization of culture, in particular the possibility for everybody to participate in art and its works.

Gernot Böhme was Professor of Philosophy at TU Darmstadt, Germany between 1977 and 2002 and has been director of the Institute for Practical Philosophy in Darmstadt since 2005. His research interests include the philosophy of science, theory of time, aesthetics, ethics, and philosophical anthropology.

Jean-Paul Thibaud, sociologist, is CNRS Senior Researcher at CRESSON/ UMR Ambiances Architectures Urbanités. His field of research covers the theory of urban ambiances, ordinary perception, and sensory ethnography of public places. In 2008 he founded the International Ambiances Network.

Ambiances, Atmospheres and Sensory Experiences of Space

Series Editors:
Rainer Kazig, CNRS Research Laboratory Ambiances – Architectures – Urbanités, Grenoble, France
Damien Masson, Université de Cergy-Pontoise, France
Paul Simpson, Plymouth University, UK

Research on ambiances and atmospheres has grown significantly in recent years in a range of disciplines, including Francophone architecture and urban studies, German research related to philosophy and aesthetics, and a growing range of Anglophone research on affective atmospheres within human geography and sociology.

This series offers a forum for research that engages with questions around ambiances and atmospheres in exploring their significances in understanding social life. Each book in the series advances some combination of theoretical understandings, practical knowledges and methodological approaches. More specifically, a range of key questions which contributions to the series seek to address includes:

- In what ways do ambiances and atmospheres play a part in the unfolding of social life in a variety of settings?
- What kinds of ethical, aesthetic, and political possibilities might be opened up and cultivated through a focus on atmospheres/ambiances?
- How do actors such as planners, architects, managers, commercial interests and public authorities actively engage with ambiances and atmospheres or seek to shape them? How might these ambiances and atmospheres be reshaped towards critical ends?
- What original forms of representations can be found today to (re)present the sensory, the atmospheric, the experiential? What sort of writing, modes of expression, or vocabulary is required? What research methodologies and practices might we employ in engaging with ambiances and atmospheres?

The Aesthetics of Atmospheres

Gernot Böhme

Edited by
Jean-Paul Thibaud

Routledge
Taylor & Francis Group

LONDON AND NEW YORK

First published 2017 by Routledge

2 Park Square, Milton Park, Abingdon, Oxfordshire OX14 4RN
711 Third Avenue, New York, NY 10017

Routledge is an imprint of the Taylor & Francis Group, an informa business

First issued in paperback 2018

Every effort has been made to contact copyright holders for their permission
to reprint material in this book. The publishers would be grateful to hear
from any copyright holder who is not here acknowledged and will undertake
to rectify any errors or omissions in future editions of this book.

British Library Cataloguing in Publication Data
A catalogue record for this book is available from the British Library

Library of Congress Cataloging in Publication Data
Names: Böhme, Gernot, author. | Thibaud, Jean-Paul, editor.
Title: The aesthetics of atmospheres / Gernot Böhme ; edited by
Jean-Paul Thibaud.
Description: New York : Routledge, 2016. | Includes index.
Identifiers: LCCN 2016010190 | ISBN 9781138688506 (hardback) |
ISBN 9781315538181 (e-book)
Subjects: LCSH: Aesthetics. | Nature (Aesthetics) | Architecture–Aesthetics. |
Light. | Sound (Philosophy)
Classification: LCC BH39 .B6172513 2016 | DDC 111/.85–dc23
LC record available at https://lccn.loc.gov/2016010190

ISBN: 978-1-138-68850-6 (hbk)
ISBN: 978-1-138-32455-8 (pbk)

Typeset in Times New Roman
by Taylor & Francis Books

Contents

vi   *Contents*

Figures

Acknowledgements from the International Ambiances Network

The International Ambiances Network (http://www.ambiances.net) aims at structuring and developing the research field of architectural and urban ambiances. It wishes to promote the sensory domain in the questioning and design of lived space. This sensitive approach of the built environment involves all the senses (sound, light, odors, touch, heat, etc.).

The International Ambiances Network favors multisensoriality and pluridisciplinarity (human and social sciences; architecture and urban planning; engineering and applied physics). It is open to a wide variety of profiles and includes research activities as well as design, teaching or artistic ones.

The International Ambiances Network wishes to thank Gernot Böhme for his trust in providing us with the opportunity to publish his book. Such an undertaking is of particular significance, given the centrality of the theory of Gernot Böhme within the thematic of sensory atmospheres. By contributing to this publication, the International Ambiances Network continues its scientific mission and work enhancement in this area. It reaffirms the importance of the dissemination and discussion of the essential texts of this field of thought and action, it calls for an open search for the diversity of languages, disciplines, and approaches, and it implements the constitution of internationally shared knowledge in the field.

Foreword

A curious paradox reflects the field of sensory atmospheres and architectural and urban ambiances. While this research domain has been growing for almost 30 years and tends increasingly to spread internationally, the seminal work of Gernot Böhme has been only partially disclosed to and published for the English-speaking public. This collection of articles intends to fill this gap by providing access of the philosophy of atmospheres developed by Gernot Böhme to a broader readership.

Gernot Böhme has gained international renown through his work on the philosophy of science and the theory of atmospheres. The present work concerns more particularly the second branch of his thought. He has devoted the past 25 years to founding and developing the idea of atmosphere, first fitting it into his ecological aesthetics of nature then thematizing it as part of his overall aesthetic. Atmosphere thus enables him to return to the prime sense of the aesthetic, in other words, conceived as a theory of sensory perception.

At the beginning of the 1990s the idea of atmosphere emerged as the touchstone of Böhme's new aesthetic. His thinking constituted an essential reference in the field, initiating the new widely shared interest in the theme of atmospheres. Of course Böhme drew on the work of leading authors to back up a project on this scale, in particular Hubertus Tellenbach's conception of the oral sense, Hermann Schmitz's philosophy of the felt body, or Walter Benjamin's concept of aura. However, it was the author of *Für eine ökologische Naturästhetik* (1989), *Atmosphäre: Essays zur neuen Ästhetik* (1995), and *Architektur und Atmosphäre* (2006) who gave its full measure to this new field of research.

The reader will certainly recognize in Böhme's aesthetic of atmospheres the underlying filiation to *Naturphilosophie*, extended here in new and original ways. His thinking is inspired by and benefits from the new phenomenology current in German thinking and the social criticism propounded in its early days by the Frankfurt school. But we should look further, paying attention to his work on Kant's *Critique of Judgment*, Plato's theory of mimesis and Aristotle's *Poetics*. Reading very widely in philosophy, Böhme established solid roots for his philosophy of atmospheres, proposing new concepts for its application. Take for instance his idea of ecstasy and the 'ecstatic' nature of things which radiates around them, or the notion of medium as the third essential term between the object perceived and the perceiving subject.

But here again his aesthetic of atmospheres cannot be seen as a purely speculative theoretical stance. It also brings into play a two-sided approach that is both critical and practical. Furthermore, Böhme has repeatedly stressed the critical value of atmosphere as an idea. Rooted in history and built into forms of social life, atmospheres are a means of revealing political issues in contemporary life and taking on board the consequences of its aestheticization. One of Böhme's recent publications – *Ästhetischer Kapitalismus* – reflects the particular importance he attaches to the political arena. Furthermore, atmospheres are not merely a matter of reception, but also of production. Their material dimension thus opens onto the world of action, architecture, and design. From this perspective, atmospheres have as much to do with *aisthesis* as with *poiesis*, witness the operational power of atmosphere as an idea, and its ability to shape and take part in the world in its making.

The present collection brings together a series of articles published over the past 25 years. Reading between the lines one can see how his aesthetic of atmospheres has evolved and gained depth, much as the intertwining arguments, which respond to and enrich one another as time passes. Turning the pages, we see how his thinking developed, matured, and reached out. Whereas the concept of atmosphere is now widely used, almost commonplace, and certainly seems almost self-evident, this collection highlights the work and invention required to gain acceptance for the idea and develop an aesthetic standpoint of such complexity. In a way the book tells us a lot about the conditions under which the idea took shape and its history.

These texts were published in various journals, in collective works and exhibition catalogues. Written in very different contexts and addressing readers of substantially different disciplinary backgrounds, they gradually map out a huge range of questions, themes, and fields covered by the idea of atmosphere. Rather than taking the articles in chronological order it seemed preferable here to regroup them by theme, giving the reader a sense of the topics of key concern to Böhme. This applies to aesthetic theory itself, with in-depth analysis of its roots and enigmas, to the aesthetic of art and nature, which opens the way for a fresh look at the question of ecology, to architecture, which reflects the operational character of atmosphere as an idea when designing and building living spaces, and the sound-and-light environments which affect and configure our daily experience. The limited number of inputs does not belie the great diversity of issues addressed, reaching from life in society to political economics, from the city to the body, from language to technology. One immediate lesson may be learned from such thematic wealth, namely the cross-the-board, transversal nature of the idea of atmosphere. It operates as a remarkable sounding board, a means of analyzing the contemporary sensory world, encompassing a broad spectrum of questions and experiences. The introduction written by Böhme specifically for this publication, demonstrates this very well, pinpointing the many artistic, social, political, and ecological issues of present concern which can be interpreted in the light of his aesthetic of atmospheres.

The publication of this collection of articles was made possible thanks to the involvement of various members of the International Ambiances Network. Our thanks to Céline Bonicco-Donato, Aurore Bonnet, Rainer Kazig, Damien Masson, Olfa Meziou, Anthony Pecqueux, Nathalie Simonnot, Daniel Siret, and Nicolas Tixier. The final correction of the book has also greatly benefited from the language skills and meticulous reading of Martin Barr and Laurent Vermeersch, whom we sincerely thank.

We also thank the publishers, editors, magazines, heads of collective works and journal issues for the authorization they gave us to republish these texts.

Jean-Paul Thibaud

Introduction

The aesthetic theory of atmospheres

Historical background

As to me, I first introduced the concept of atmosphere in my German book *Towards an Ecological Aesthetics*.[1] The point of the book was a critique of scientific ecology and a plea to introduce the human factor into the science of environment. Our main interest, I argued, was not in the natural inter-relatedness of nature as such, but in our own environment, i.e. in human beings. This interest on the general scope must lead to a social–natural science.[2] But the main concern of the book was to introduce the aesthetic perspective into the science of ecology: what affects human beings in their environment are not only just natural factors but also aesthetic ones.

If you do not feel well in an environment, the reason might not be a toxic agent in the air but aesthetic impressions.

For example: again and again the population of my home town, the city of Darmstadt, complained saying "There is a bad smell in the air." The origin supposed was the production site of Merck, a big chemical and pharmaceutical company. Well, the scientists of Merck made an investigation the outcome of which was: no toxic substances in the air. No toxins, no problem. But there was a problem: the inhabitants of Darmstadt "did not feel well."

This "feeling well or not" in a certain environment clearly is an indicator of the aesthetic qualities of it. This is the point where aesthetics come into ecology. The elements of the environment are not only causal factors which affect human beings as organisms but they produce an impression on their feeling (*Befindlichkeit*). And what mediates objective factors of the environment with aesthetic feelings of a human being is what we call *atmosphere*. The atmosphere of a certain environment is responsible for the way we feel about ourselves in that environment.

Atmosphere is what relates objective factors and constellations of the environment with my bodily feeling in that environment. This means: atmosphere is what is *in between*, what mediates the two sides. Two main traits of the theory of atmospheres arise from this. Namely, first, that atmosphere is something in between subject and object and can therefore be approached in two different ways: either from a perception aesthetics or a production aesthetics viewpoint. Atmospheres are quasi-objective, namely they are out there; you can enter an atmosphere and you can be surprisingly caught by an atmosphere. But on the other hand atmospheres are not beings like things; they are nothing without a subject feeling them. They are subjective facts in the sense of Hermann Schmitz: to talk about atmospheres, you must characterize them by the way they affect you. They tend to bring you into a certain mood, and the way you name them is by the character of that mood. The atmosphere of a room may be oppressive, the atmosphere of a valley may be joyful. But on the other side you can argue about atmospheres and you even can agree with others about what sort of atmosphere is present in a certain room or landscape. Thus atmospheres are quasi-objective or something existent intersubjectively.

But, as mentioned, you can approach the phenomenon of atmospheres not only from the side of perception aesthetics but also from that of productions aesthetics. This is why stage design is a kind of a paradigm for the whole theory and practice of atmospheres: you can learn from a stage designer what means are necessary in order to produce a certain climate or atmosphere on the stage: what the sound should be like, how the stage is illuminated, what materials, colors, objects, signs should be used, and in what way should the space of the stage itself be arranged. The art of stage setting again proves that atmospheres are something quasi-objective. Namely, if each member of the audience were to perceive the climate of the stage in a different way, the whole endeavor of stage setting would be useless.

The origin of the term atmosphere *and its original use as a concept* in science and humanities

The term *atmosphere* was originally used within meteorological contexts. Here it designated the upper part of the air mantling the earth. But since the eighteenth century *atmosphere* was used as metaphor describing a certain mood hanging in the air. The mediating link obviously is the weather: the weather is affecting my mood – a rising thunderstorm may frighten me, bright weather may raise my spirits.

Today atmosphere may be defined briefly as *tuned space*, i.e. a space with a certain mood. From here two more traits of the theory of atmospheres can be advanced: atmospheres are always something spatial, and atmospheres are always something emotional.

We talk about atmospheres by naming their characteristics. These are their tendencies to modify my own mood. The serious atmosphere of a gathering may make me serious; the melancholic atmosphere of garden scenery may make me melancholic.

The first scientific use of the term *atmosphere* in this sense is to be found in Hubert Tellenbach's book *Geschmack und Atmosphäre*.[3] This book, which actually deals with the sphere of the oral, uses the term in particular for the smell of the nest: atmosphere is what makes you feel at home. This book is of lasting value for the theory of atmospheres because it links the natural with the cultural realm. Atmosphere is something which affects us deeply, that means on the level of bodily feeling.

Later the concept of atmosphere was elaborated by the so-called new phenomenology, in particular by its founder Hermann Schmitz.[4] He conceives of atmospheres as being overwhelming emotional powers, or – as he sees it – quasi-objective feelings. He was influenced by the research on the numinous as carried out by Rudolf Otto.[5]

Applications of the concept of atmospheres

Scenography

We mentioned already the art of stage setting could be used as a paradigm for the theory of atmosphere. Here, long before anybody thought of atmospheres, a practice of soliciting atmospheres was developed: stage setters knew how to produce a certain mood, or – as they call it – a certain climate on the stage. So, what can to be learned from the tradition of stage setting is:

- Atmospheres can be produced.
- Atmospheres are something out there, quasi-objective.
- Atmospheres are produced by certain agents or factors, in particular by sound and illumination, but also by the geometry of a room, by signs, pictures, etc.

But the art of stage setting is sort of tacit knowledge; you would be hard pressed to find a book telling you how and by what means a certain atmosphere can be to produced. This is why a book seemingly from a quite different strand must be mentioned, namely C. C. L. Hirschfield's theory of English gardening.[6] This book obviously is inspired by the world of the theater. Thus Hirschfield talks about natural scenery and of the emotional character of it – what we would refer to as its atmosphere. But what is important is that he gives detailed instruction as to what sort of trees and other plants a certain mood may produce, how the light falling through the leaf must be, how the murmur of the brooks, whether the sight must be open or closed; in short, he talks about atmospheres like an artisan who knows to *make* them.

This leads us to an extension of the field: stage design might be a useful paradigm of producing atmospheres, but today it is much better to talk of scenography. Under this very old term[7] a new discipline is developing, the job of which is *staging* of everything: this might be political, sportive, or cultural events. The point is that these human practices are no longer performed, as such, but must be set in scene, performed in a certain frame, celebrated in a way. Politics before the camera, sportive competitions as a festival; and the presentation of artworks must take place within a certain setting, a certain arrangement and illumination. One of the earliest fields of this type of extended scenography is the staging of commodities. The origin of this was located by Walter Benjamin within the arcades of nineteenth-century Paris. Today it is not only the single commodity what is on stage, but the *brand* must be staged, if possible presented as a whole world. The Nike-World is an example of this endeavor, but other brands like Joop or Dior might be even more extended, covering a strand of commodities far beyond the original.

Commodity aesthetics

The concept of commodity aesthetics was introduced by Wolfgang Fritz Haug.[8] But his book concentrated particularly on the packaging of commodities, how they were presented in the marketplace. Since his time, we have seen an extension of commodity aesthetics into the fields of production and consumption.

The most impressive example of the first is Volkswagen's production of the Phaeton automobile in Dresden. In a corner of the Great Baroque Garden they built the so-called Gläserne Manufaktur – this may be translated as "production site in the shop window display."[9] In this huge glass building visitors can watch how the Phaeton is finished in the assembly line. The whole process is celebrated in a glamorous environment like a church ceremony.

The other field of extended commodity aesthetics is consumption. Whereas in Haug's book the aesthetics of the commodity is its packaging, which is soon discarded, now we notice that the aesthetic outfit of the commodity has a function in the realm of consumption. The first step in this direction was noticed by Jean Baudrillard:[10] the commodity got a function as a status symbol. Today many commodities are not really used in a literal sense but they get their use-value merely as ingredients of a certain lifestyle of the user.

This development was the reason why I began talking about an aesthetic economy.[11] Commodities are valued in the aesthetic economy where they now merely satisfy basic needs; for their staging-value, they are valued to the extent that they help individuals or groups to stage their lifestyles. Here commodities have their use-value; a means of producing a certain atmosphere. This gives us a reason to talk about a new type of commodity value besides the Marxian use-value and exchange-value, namely to attribute the new type of stage-value to commodities.

Advertising

This new use of commodities, namely as a means to produce an atmosphere, caused a shift in advertising. Whereas traditional advertising, say, from the nineteenth century up to the first half of the twentieth century represented commodities as well made and useful, contemporary advertising does not present the commodities as such but rather a scene within which they have a certain function, namely contributing to an atmosphere. So you might notice a bottle of Beck's beer in the hand of a member of a sailing crew, or a Vuitton bag in an outdoor picnic scene. The appeal of advertising is not to a customer who wants to make use of a commodity but to somebody who wants to be embedded into a certain atmosphere of life. This also means customers want to belong to a certain group; they want to distinguish themselves from the crowd by association with a certain lifestyle. Thus the aesthetics of atmospheres in advertising means that commodities are not presented as things which are useful within a certain practice but as signs which help to produce a certain atmosphere in life.

Architecture and design

One of the main applications of the aesthetic theory of atmospheres is architecture and design. Architecture and design have always produced atmospheres, but the thinking about architecture mainly concentrated on buildings and their visual representation; and thinking about design concentrated on the form or shape of things. This type of thinking came to its peak with Bauhaus modernity and found adequate expression in the slogan "form follows function." But since the turn to postmodernity we have a new humanism in both fields and that means that the way we experience buildings and the surroundings, how we feel as visitors or people who live there, comes to the fore. In the theory of design, the situation is comparable: it is not only the function or, say, the use-value of things which is at stake, but what sort of impression the objects make. It is necessary to observe that this turn has something to do with the transformation of capitalism into an aesthetic economy.

But what interests us here is the shift in thinking both in architecture and design as a consequence of the theory of atmospheres. We said: atmospheres are something spatial and at the same time something emotional. If you are explicitly considering atmospheres in architecture and city planning the main topic of your considerations is space. Architecture is not just about buildings but essentially about spaces. Architecture is opening and closing spaces, it sets points of concentration and therefore of orientation in space; it determines directions, it frames outlooks. And all this for people visiting or dwelling there. That means that the way people feel in rooms and spaces, how they move around, how they can follow bodies and lines of buildings is the main point of interest.

The situation is comparable in the art of design. Here a shift of consideration took place, which again is determined by the perspective of the customer. Whereas in traditional theory of design one was talking about the shape and the properties of things, it is now about "ecstasies." I use the Greek word *ecstasies* to indicate the way things are radiating into space and thus contributing to the formation of an atmosphere. *Ecstatics* is the way things make a certain impression on us and thus modifying our mood, the way we feel ourselves.

In the fields of architecture and design the turn is from the form or shape of things to their contribution of tuning the space of our bodily presence.

Art

The impact of the aesthetic theory of atmospheres to art is primarily a shift in perception aesthetics. It is well known that since Kant and in particular following Hegel aesthetics became a theory of judgment – judgment on and about works of art. This meant that aesthetics primarily was useful for the educated elite and for art critics. The consequence was that guidance in art exhibitions means information about the artist, his technique, his time – but on the other side the guided visitor has no real chance to make observations and experiences of his own. This is the more regrettable because the interest in art has widened far beyond the educated elite. Now, the theory of atmospheres opens a quite different approach to works of art, i.e. an approach which is not guided by art history, iconography, and semantics. The main goal of visiting an exhibition is not learning or information but having experiences. Guidance no longer means information but assistance in approaching the work of art and in preparation of one's own experiences.[12]

This turn from meaning to experience in the perception of works of art is met by a certain development in art itself. There are quite a few paintings which have no meaning, in particular monochromic painting, but the whole movement of abstract expressionism may be mentioned here. More explicit as to the requirement to have experiences – and the means of being bodily present at the place where the work of art is – is land-art and the art of sound installations. These types of artwork are on the one hand explicitly related to *their* place and on the other they are ephemeral.[13] The consequence is that in order to adequately appreciate what these works of art are requires exposing oneself to the atmosphere they are radiating.

Conclusion

The *detection* of atmospheres was a great step forward for philosophy: dedicated to the clear and distinct – at least since Descartes – philosophy for the first time came to conceive of and to talk about a vague and rather subjective phenomenon. The phenomenological analysis of atmospheres was very fruitful and prepared the ground for the theory to be applied in many fields. The most important – and successful – application was within aesthetics. The material background of this success may be seen in the ubiquitous aestheticization of our lifeworld or – taking it more from the side of production – of staging of everything, every event and performance.

The theory of atmospheres becoming an aesthetic theory thus turned out to be a critical theory of our contemporary civilization. It reveals the theatrical, not to say manipulative character of politics, commerce, of the event-society.[14] This for the critical power of the theory. But taking it as a positive theory of certain phenomena it opened up a lot of new perspectives for architecture, design, and art. It made the spatial and the experience of space and places a main subject and hence rehabilitated the ephemeral in arts. Taking the numerous impacts in many fields together it initiated a new humanism: the individual as a living person and his or her perspective being taken seriously in architecture and design; and it fosters the ongoing *democratization* of culture, in particular the possibility for everybody being able to participate in art and its works.

Notes

1 Gernot Böhme, *Für eine ökologische Naturästhetik*, Frankfurt/M., Suhrkamp, 3rd edn, 1999 [1989].
2 Gernot Böhme and Engelbert Schramm (eds.), *Soziale Naturwissenschaft. Wege zur Erweiterung der Ökologie*, Frankfurt/M., Fischer, 1985.
3 Hubertus Tellenbach, *Geschmack und Atmosphäre. Medien menschlichen Elementarkontaktes*. Salzburg, Otto Müller Verlag, 1968.
4 Hermann Schmitz, *System der Philosophie*. Bonn, Bouvier, 1964, Bd. III.1 Die Wahrnehmung.
5 Rudolf Otto, *Das Heilige: Über das Irrationale in der Idee des Göttlichen und sein Verhältnis zum Rationalen*. Breslau, Trewendt & Granier, 1917; Nachdruck, München, Beck, 2004.
6 C. C. L. Hirschfeld, *Theorie der Gartenkunst*, Leipzig, 5 Bde., 1779–85.
7 Aristotle says that Sophocles already practiced *skenographia, Poetics*, 1449a18.
8 Wolfgang Fritz Haug, *Kritik der Warenästhetik*, Frankfurt/M., Suhrkamp, 1971.
9 See my article "Fortschritte der Warenästhetik. Passagen an den Rändern der Kulturwissenschaft," in N. Adamowsky, P. Matussek (Hrsg.) *Auslassungen. Leerstellen als Movens der Kulturwissenschaft*. Würzburg, Königshausen & Neumann, 2004, S. 31–8.
10 Jean Baudrillard, *For a Critique of the Political Economy of the Sign*, trans. C. Levin, St Louis, Telos Press, 1981 [1972].
11 Gernot Böhme, "Contribution to the Critique of Aesthetic Economy," *Thesis Eleven*, 73, May 2003, 71–82. Reprinted in this book, as Chapter 7.
12 One of the most gifted art guides in this sense was Michael Bockemühl. See his *Die Wirklichkeit des Bildes. Bildrezeption als Bildproduktion – Rothko, Newman, Rembrandt, Raphael* [Habilitationsschrift], Urachhaus Verlag, Stuttgart, 1985; and J. M. W. Turner, 1775–1851 – *The World of Light and Colour*, Cologne, 2000.
13 See Gernot Böhme, *Die sanfte Kunst des Ephemeren*. Essen, Verlag der fadbk, 2008, in M. Fliescher, F. Goppelsröder, and D. Mersch (Hrsg.), *Sichtbarkeiten I. Erscheinen. Zur Praxis des Präsentiven*. Berlin, Diaphanes, 2013, 87–108.
14 See Gerhard Schulze, *Die Erlebnisgesellschaft: Kultursoziologie der Gegenwart*. Frankfurt/M., Campus, 1992.

Part I
Theory: aesthetics and aesthetical economy

1 Atmosphere as a fundamental concept of a new aesthetics[1]

Atmosphere

The expression "atmosphere" is not foreign to aesthetic discourse. On the contrary, it occurs frequently, almost of necessity in speeches at the opening of exhibitions, in art catalogues, and in eulogies in the form of references to the powerful atmosphere of a work, to atmospheric effect, or a rather atmospheric mode of presentation. One has the impression that "atmosphere" is meant to indicate something indeterminate, difficult to express, even if it is only in order to hide the speaker's own speechlessness. It is almost like Adorno's "more," which also points in evocative fashion to something beyond rational explanation and with an emphasis which suggests that only there is the essential, the aesthetically relevant to be found.

This use of the word "atmosphere" in aesthetic texts, oscillating between embarrassment and emphasis, corresponds to its use in political discourse. Here too everything apparently depends on the atmosphere in which something occurs and where the improvement of the political atmosphere is the most important thing. On the other hand, the report that negotiations took place "in a good atmosphere" or led to an improvement in the atmosphere is only the euphemistic version of the fact that nothing resulted from a meeting. This vague use of the expression atmosphere in aesthetic and political discourse derives from a use in everyday speech which is in many respects much more exact. Here the expression "atmospheric" is applied to persons, spaces, and to nature. Thus one speaks of the serene atmosphere of a spring morning or the homely atmosphere of a garden. On entering a room one can feel oneself enveloped by a friendly atmosphere or caught up in a tense atmosphere. We can say of a person that s/he radiates an atmosphere which implies respect, of a man or a woman that an erotic atmosphere surrounds them. Here too atmosphere indicates something that is in a certain sense indeterminate, diffuse but precisely not indeterminate in relation to its character. On the contrary, we have at our disposal a rich vocabulary with which to characterize atmospheres, that is, serene, melancholic, oppressive, uplifting, commanding, inviting, erotic, etc. Atmospheres are indeterminate, above all as regards their ontological status. We are not sure whether we should attribute them to the objects or environments from which they proceed or to the subjects who experience them. We are also unsure where they are. They seem to fill the space with a certain tone of feeling like a haze.

The frequent, rather embarrassed use of the expression atmosphere in aesthetic discourse leads one to conclude that it refers to something which is aesthetically relevant but whose elaboration and articulation remains to be worked out. As my introductory remarks suggest, the introduction of "atmosphere" as a concept into aesthetics should link up with the everyday distinctions between atmospheres of different character. Atmosphere can only become a concept, however, if we succeed in accounting for the peculiar intermediary status of atmospheres between subject and object.

A new aesthetics

I first made the call for a new aesthetics[2] in my book *Für eine ökologische Naturästhetik* (1989). This call has been misunderstood as fundamental ecology[3] or as organicism.[4] It is true that one aim of my book was the introduction of aesthetic viewpoints into ecology. It is true that in this book what we perceive is also called a form of nourishment, and that aesthetic nature remains our goal.[5] The call, however, goes much further. I quoted Goethe in order to recall that "it makes a great difference from which side one approaches a body of knowledge, a science, through which gate one gains access." Aesthetics opens up as a completely different field if it is approached from ecology, something completely different from its tradition of presentation from Kant up to Adorno and Lyotard. The quest for an aesthetics of nature as an aesthetic theory of nature requires that we reformulate the theme of aesthetics as such. The new resulting aesthetics is concerned with the relation between environmental qualities and human states. This "and," this in-between, by means of which environmental qualities and states are related, is atmosphere. What is new in this new aesthetics can be formulated in threefold form.

1   The old aesthetics is essentially a judgmental aesthetics, that is, it is concerned not so much with experience, especially sensuous experience – as the expression "aesthetics" in its derivation from the Greek would suggest – as with judgments, discussion, conversation. It may have been the case that the question of taste and individual affective participation (under the title *Faculty of approval*) in a work of art or in nature provided the original motive for aesthetics. With Kant at the latest, however, it became a question of judgment, that is, the question of the justification for a positive or negative response to something. Since then the social function of aesthetic theory has been to facilitate conversation about works of art. It supplies the vocabulary for art history and art criticism, for the speeches at exhibitions and prize-givings and for essays in catalogues. Sensuousness and nature have in this fashion disappeared from aesthetics.

2   The central place of judgment in aesthetics and in its orientation to communication led to a dominance of language and to the present dominance of semiotics in aesthetic theory. This situation gives literature precedence over the other kinds of art, which are also interpreted by means of the schema of language and communication. Aesthetics can be presented under the general heading "languages of art."[6] It is not, however, self-evident that an artist intends to communicate something to a possible recipient or observer. Neither is it self-evident that a work of art is a sign, insofar as a sign always refers to something other than itself, that is, its meaning. Not every work of art has a meaning. On the contrary, it is necessary to remember that a work of art is first of all itself something, which possesses its own reality. This can be seen in the contortions that semiotics engages in with the concept of the "iconic sign" in order to be able to subsume paintings under the sign. Iconic signs do not reproduce the object but "some conditions of the perception of the object."[7] Through this use a painting of Mr. Smith is to be understood as a sign for Mr. Smith, even though it is in a certain way Mr. Smith: "That is Mr. Smith" is the answer to the question "Who is that?" Thus, for example, Eco declares *Mona Lisa* to be an iconic sign for Mona Lisa. Apart from the fact that the relation of the picture *Mona Lisa* to a person Mona Lisa is highly questionable, as Gombrich has shown in his essay on the portrait,[8] nobody understands by "Mona Lisa" the person Mona Lisa but the painting and it is this which is experienced. The painting does not refer to its meaning as a sign (a meaning which could only be thought); the painting is in a certain sense what it itself represents, that is, the represented is present in and through the painting. Of course, we can also read and interpret such a painting but this means cutting out or even denying the experience of the presence of the represented, namely the atmosphere of the painting.[9]

3   After its original orientation, aesthetics very quickly became a theory of the arts and of the work of art. This, together with the social function of aesthetics as background knowledge for art criticism, led to a strongly normative orientation: it was not a question of art but of real, true, high art, of the authentic work of art, the work of art of distinction. Although aestheticians were fully aware that aesthetic work is a much broader phenomenon, it was registered at best only marginally and disdainfully, namely as mere beautification, as craftsmanship, as kitsch, as useful or applied art. All aesthetic production was seen from the perspective of art and its measure. Walter Benjamin introduced a change of perspective with his essay *The Work of Art in the Age of Its Technical Reproduction*.[10] On the one hand the possibility of Pop Art was envisaged before it actually existed, while on the other the aestheticization of the life world was thematized as a serious phenomenon under the formula of the "aestheticization of politics." The primary task of aesthetics is no longer to determine what art is and to provide means for art criticism. Rather the

theme of aesthetics is now the full range of aesthetic work, which is defined generally as the production of atmospheres and thus extends from cosmetics, advertising, interior decoration, stage sets to art in the narrower sense. Autonomous art is understood in this context as only a special form of aesthetic work, which also has its social function, namely the mediation of the encounter and response to atmospheres in situations (museums, exhibitions) set apart from action contexts.

The new aesthetics is thus as regards the producers a general theory of aesthetic work, understood as the production of atmospheres. As regards reception it is a theory of perception in the full sense of the term, in which perception is understood as the experience of the presence of persons, objects, and environments.

Benjamin's aura

"Atmosphere" is an expression which occurs frequently in aesthetic discourse but is not up to now a concept of aesthetic theory. Nevertheless, there is a concept which is, so to speak, its substitute representative in theory – the concept of aura, introduced by Benjamin in his essay *The Work of Art*. Benjamin sought through the concept of aura to determine that atmosphere of distance and respect surrounding original works of art. He hoped thereby to be able to indicate the difference between an original and its reproductions and thought that he could define a general development of art through the loss of aura, which was brought about by the introduction of technical means of reproduction into art production. In fact, the artistic avant-garde sought to expel the aura of art through the reunion of art and life. Duchamp's ready-mades, Brecht's disillusioning of the theatre, and the opening up of art to Pop Art are examples. They failed or their outcome is at least paradoxical. The very fact that Duchamp declared a ready-made to be a work of art lent it aura and now his ready-mades display in museums as much distance and command the same respect as a sculpture by Veit Stoss. The avant-garde did not succeed in discarding aura like a coat, leaving behind them the sacred halls of art for life. What they did succeed in doing was to thematize the aura of artworks, their halo, their atmosphere, their nimbus. And this made it clear that what makes a work an artwork cannot be grasped solely through its concrete qualities. But what exceeds them, this "more," the aura, remained completely undetermined. "Aura" signifies as it were atmosphere as such, the empty characterless envelope of its presence.

Nevertheless, it is worth holding on to what is already implied in Benjamin's concept of aura for the development of the concept of atmosphere as a fundamental concept of aesthetics. The genesis of aura is paradoxical; Benjamin introduced it to characterize works of art as such. He derives it, however, from a concept of nature. I quote the whole passage on account of the special significance of this genesis:

What is aura actually? A strange tissue of space and time: unique appearance of distance, however near it may be. Resting on a summer evening and following a mountain chain on the horizon or a branch, which throws its shadow on the person at rest – that is to breathe the aura of these mountains or this branch. With this definition it is easy to comprehend the particular social determination of the present decay of aura.[11]

When Walter Benjamin speaks of the "appearance" of distance, he does not mean that distance appears; rather, he is speaking of the phenomenon of distance which can also be discerned in things which are close. This is the unattainability and distance which is discernible in works of art. He has already introduced the "unique" and commits a *petitio principii*, since it is precisely through aura that the uniqueness of artworks is to manifest itself. The aura itself is not unique, it is repeatable. Let us now consider the experience from which the concept of aura derives. The examples show that Benjamin posits for the experience of aura first a certain natural impression or mood as background and second a certain receptivity in the observer. Aura appears in the situation of ease, that is, observation, in a physically relaxed and work-free situation. Following Hermann Schmitz, we could say that "summer afternoon" and "resting" – Benjamin's example suggests that he observes mountain chain and branch lying on his back – imply a bodily tendency to privatize experience. The aura can now appear in relation to a distant mountain chain, the horizon, or a branch. It appears in natural objects. Aura proceeds from them, if the observer lets them and himself be, that is, refrains from an active intervention in the world. And aura is clearly something which flows forth spatially, almost something like a breath or a haze – precisely an atmosphere. Benjamin says that one "breathes" the aura. This breathing means that it is absorbed bodily, that it enters the bodily economy of tension and expansion, that one allows this atmosphere to permeate the self. Precisely this dimension of naturalness and corporeality in the experience of aura disappears in Benjamin's further use of the expression, although in this first version his exemplary presentation of the experience of aura serves as its definition.

We retain the following: something like aura according to Benjamin is perceptible not only in art's proudest or original works. To perceive aura is to absorb it into one's own bodily state of being. What is perceived is an indeterminate spatially extended quality of feeling. These considerations serve to prepare us for the elaboration of the concept of atmosphere in the framework of Hermann Schmitz's philosophy of the body.

The concept of atmosphere in the philosophy of Hermann Schmitz

When we stated above that "atmosphere" is used as an expression for something vague, this does not necessarily mean that the meaning of this expression is itself vague. Admittedly, it is difficult, owing to the peculiar intermediary position of the phenomenon between subject and object, to determine the status of atmospheres and thereby transform the everyday use of atmospheres into a legitimate concept. In raising the claim that atmosphere constitutes the fundamental concept of a new aesthetics, it is not necessary to establish the legitimacy of this concept, since Hermann Schmitz's philosophy of the body already provides an elaboration of the concept of atmospheres. Schmitz's concept of atmospheres has a precursor in Ludwig Klages's idea of the "reality of images." In his early work *Vom kosmogonischen Eros* Klages set out to show that appearances (images) possess in relation to their sources a relatively independent reality and power of influence. This thesis of the relative independence of images derives in part from the disappointing experience that the physiognomy of a person can hold a promise which is not fulfilled.[12] Klages thus conceives an "eros of distance," which unlike the Platonic Eros does not desire closeness and possession but keeps its distance and is fulfilled by contemplative participation in the beautiful. Images in this sense have reality in that they can take possession of a soul. Klages developed these insights systematically in his *Grundlegung der Wissenschaft vom Ausdruck*[13] (as well as in *Der Geist als Widersacher der Seele*). What was termed the reality of images is now treated under the headings expression, appearance, character, and essence. It is important to note here that these expressive qualities, especially those of living being, are accorded a kind of self-activity. "The expression of a state of being is composed in such a way that its appearance can call forth the [corresponding] state."[14] Expressive appearances are powers of feeling and are therefore called at times daemons or even souls. The perceiving soul by contrast has a passive role: perception is affective sympathy. In his concept of atmosphere Schmitz takes over two aspects of Klages's idea of the reality of images: on the one hand their relative independence in relation to things, and on the other their role as active instances of feeling which press in from outside the affective power.

Schmitz's concept of atmosphere uncouples the phenomenon in question even further from things: as he no longer speaks of images, physiognomy plays no role. In its place he develops the spatial character of atmospheres. Atmospheres are always spatially "without borders, disseminated and yet without place that is, not localizable." They are affective powers of feeling, spatial bearers of moods.

Schmitz introduces atmospheres phenomenologically, that is, not through definition but through reference to everyday experiences such as those indicated above, the experience of a strained atmosphere in a room, of an oppressive thundery atmosphere, or of the serene atmosphere of a garden. The legitimacy of this use of atmospheres derives for Schmitz on the one hand from the phenomenological method which recognizes what is indisputably given in experience as real, and on the other hand from the context of his philosophy of the body. His philosophy of the body removes – at least partially – the insecure status of atmospheres, which we registered above against the background of the subject–object dichotomy. According to this dichotomy, atmospheres, if we accept their relative or complete independence from objects, must belong to the subject. And in fact this is what happens when we regard the serenity of a valley or the melancholy of an evening as projections, that is, as the projection of moods, understood as internal psychic states. This conception is certainly counter-phenomenal in cases where the serenity of the valley or the melancholy of the evening strike us when we are in a quite different mood and we find ourselves seized by these atmospheres and even correspondingly changed. Within the frame of his historical anthropology Schmitz shows that the projection thesis supposes a foregoing introjection. He shows how early in our culture, that is, in the Homeric period, feelings were experienced as something "outside," as forces which actively intrude into the human body. (This is Schmitz's reconstruction of the Greek world of the gods). Against this background something like the soul appears as a "counter-phenomenal construction." What is phenomenally given, that is, sensed, is the human body in its economy of tension and expansion and in its affectivity which manifests itself in bodily impulses. Schmitz can thus define feelings as follows: they are "unlocalized, poured forth atmospheres ... which visit (haunt) the body which receives them ... affectively, which takes the form of ... emotion."[15]

We can see here the possibility of a new aesthetics which overcomes not only the intellectualism of classical aesthetics but also its restriction to art and to phenomena of communication. Atmospheres are evidently what are experienced in bodily presence in relation to persons and things or in spaces. We also find in Schmitz the beginnings of aesthetics, but one which draws only hesitantly on the potential of the concept of atmosphere. The initial steps are to be found in volume III, 4 of his system of philosophy. He remains traditional in that he does not abandon the restriction of aesthetics to art. Aesthetics appears as the subparagraph of the article art: the aesthetic sphere presupposes an "aesthetic attitude," that is, an attitude which permits the distanced influence of atmospheres. This attitude presupposes on the one hand the cultivation of the aesthetic subject, and on the other hand the "artistic setting," that is, gallery and museum outside the sphere of action. Schmitz's approach suffers above all from the fact that he credits atmospheres with too great an independence from things. They float free like gods and have as such nothing to do with things, let alone being their product. At most, objects can capture atmospheres, which then adhere to them as a nimbus. In fact (for Schmitz), the independence of atmospheres is so great and the idea that atmospheres proceed from things so distant that he regards things as aesthetic creations (*Gebilde*) if atmospheres impress their stamp upon them. He then defines aesthetic creations as follows:

> A sensuous object of a lower degree (for example, thing, sound, scent, color) I designate *aesthetic creations* if in this way they absorb into themselves atmospheres, which are objective feelings, in a quasi-corporeal fashion and thereby indicate a corporeal emotion through them.[16]

The impression or coloring of a thing through atmospheres must be interpreted according to Schmitz by means of the classical subjectivist "as-if formula." That is to say, we designate a valley as serene because it appears as if it is imbued with serenity.

The strength of Schmitz's approach, which is a quasi-aesthetics of reception, in that he can account for perception in the full sense as affective impression by atmospheres, is countered by its weakness in terms of an aesthetics of production. His conception of atmospheres rules out the possibility that they could be produced by qualities of things. This means that the whole sphere of aesthetic work is excluded from the perspective of his approach.

The thing and its ecstasies

In order to legitimate the idea of atmospheres and overcome their ontological unlocalizability, it is necessary to liberate them from the subjective–objective dichotomy. Schmitz's philosophy of the body shows that profound changes of thought are required on the side of the subject. We must abandon the idea of the soul in order to undo the "introjections of feelings," and the human being must be conceived essentially as body, such that his/her self-givenness and sense of self are originally spatial: to be bodily self-aware means at the same time the awareness of my state of being in an environment – how I feel here.

The same is now required for the object side. The difficulty here of forming a legitimate concept of atmospheres lies in the classical ontology of the thing, which cannot be fully developed and analyzed here. The decisive point is this: the qualities of a thing are thought of as "determinations." The form, color, even the smell of a thing is thought of as that which distinguishes it, separates it off from outside and gives it its internal unity. In short: the thing is usually conceived in terms of its closure. It is extremely rare that a philosopher emphasizes, as Isaac Newton for instance does, that perceptibility belongs essentially to the thing. Ontological counter-conceptions, such as that of Jakob Böhme, who conceives of things according to the model of a musical instrument, exist only as a crypto-tradition. The dominant conception on the contrary is that formulated by Kant, that it is possible to *think* a thing with all its determinations and then pose the question whether this completely determined thing actually exists. It is obvious what a hostile hindrance such a way of thinking presents for aesthetics. A thing is in this view what it is, independent of its existence, which is ascribed to it ultimately by the cognitive subject, who "posits" the thing. Let me illustrate. If we say for example: a cup is blue, then we think of a thing which is determined by the color blue which distinguishes it from other things. This color is something which the cup "has." In addition to its blueness we can also ask whether such a cup exists. Its existence is then determined through a localization in space and time. The blueness of the cup, however, can be thought of in quite another way, namely as the way, or better, a way, in which the cup is present in space and makes its presence perceptible. The blueness of the cup is then thought of not as something which is restricted in some way to the cup and adheres to it, but on the contrary as something which radiates out to the environment of the cup, coloring, or "tincturing" in a certain way this environment, as Jakob Böhme would say. The existence of the cup is already contained in this conception of the quality "blue," since the blueness is a way of the cup being there, an articulation of its presence, the way or manner of its presence. In this way the thing is not thought of in terms of its difference from other things, its separation and unity, but in the ways in which it goes forth from itself. I have introduced for these ways of going forth the expression "the ecstasies of the thing."

It should not cause difficulty to think of colors, smells, and how a thing is tuned as ecstasies. This is already apparent in the fact that in the classical subject–object dichotomy they are designated as "secondary qualities," that is, as qualities which do not in themselves belong to the thing except in relation to a subject. What is also required, however, is to think of so-called primary qualities such as extension and form as ecstasies. In the classical ontology of the thing, form is thought of as something limiting and enclosing, as that which encloses inwardly the volume of the thing and outwardly limits it. The form of a thing, however, also exerts an external effect. It radiates as it were into the environment, takes away the homogeneity of the surrounding space and fills it with tensions and suggestions of movement. In the classical ontology the property of a thing was thought to be its occupation of a specific space and its resistance to other things entering this space. The extension and volume of a thing, however, are also externally perceptible; they give the space of its presence weight and orientation. The volume, that is, the voluminosity of a thing is the power of its presence in space.

On the basis of an ontology of the thing changed in this fashion, it is possible to conceive atmospheres in a meaningful way. They are spaces insofar as they are "tinctured" through the presence of things, of persons, or environmental constellations, that is, through their ecstasies. They are themselves spheres of the presence of something, their reality in space. As opposed to Schmitz's approach, atmospheres are thus conceived not as free floating but on the contrary as something that proceeds from and is created by things, persons, or their constellations. Conceived in this fashion, atmospheres are neither something objective, that is, qualities possessed by things, and yet they are something thing-like, belonging to the thing in that things articulate their presence through qualities – conceived as ecstasies. Nor are atmospheres something subjective, for example, determinations of a psychic state. And yet they are subject-like, belong to subjects in that they are sensed in bodily presence by human beings and this sensing is at the same time a bodily state of being of subjects in space.

It is immediately evident that this changed ontology of the thing is favorable to aesthetic theory, that it amounts to its liberation. Aesthetic work in all its dimensions comes into view. Even in the narrower sphere of art, for example, the fine arts, one can see that, precisely speaking, an artist is not concerned with giving a thing – whether a block of marble or a canvas – certain qualities, formed, or colored in such and such a fashion, but in allowing it to go forth from itself in a certain fashion and thereby make the presence of something perceptible.

The making of atmospheres

Atmosphere designates both the fundamental concept of a new aesthetics and its central object of cognition. Atmosphere is the common reality of the perceiver and the perceived. It is the reality of the perceived as the sphere of its presence and the reality of the perceiver, insofar as in sensing the atmosphere s/he is bodily present in a certain way. This synthetic function of atmosphere is at the same time the legitimation of the particular forms of speech in which an evening is called melancholy or a garden serene. If we consider it more exactly, such a manner of speech is as legitimate as calling a leaf green. A leaf does have the objective property of being green. A leaf equally can only be called green insofar as it shares a reality with a perceiver. Strictly speaking, expressions such as "serene" or "green" refer to this common reality, which can be named either from the side of the object or from the side of the perceiver. A valley is thus not called serene because it is in some way similar to a cheerful person but because the atmosphere which it radiates is serene and can put this person into a serene mood.

This is an example of how the concept of atmosphere can clarify relations and render intelligible manners of speech. But what do we know about atmospheres? Classical aesthetics dealt practically only with three or four atmospheres, for example, the beautiful, the sublime – and then the characterless atmosphere or "atmosphere as such," aura. That these themes involve a question of atmospheres was, of course, not clear and many investigations will have to be read again or rewritten. Above all the extraordinary limitation of the previous aesthetics now becomes evident, since there are in fact many more atmospheres, not to say infinitely many: serene, serious, terrifying, oppressive, the atmosphere of dread, of power, of the saint, and the reprobate. The multiplicities of the linguistic expressions which are at our disposal indicate that a much more complex knowledge of atmospheres exists than aesthetic theory suggests. In particular, we may presume an extraordinarily rich wealth of knowledge of atmospheres in the practical knowledge of aesthetic workers. This knowledge must be able to give us insight into the connection between the concrete properties of objects (everyday objects, artworks, natural elements) and the atmosphere which they radiate. This perspective corresponds approximately with the question in classical aesthetics as to how the concrete properties of a thing are connected with its beauty, except that now the concrete properties are read as the ecstasies of the thing and beauty as the manner of its presence. Aesthetic work consists of giving things, environments, or also the human being such properties from which something can proceed. That is, it is a question of "making" atmospheres through work on an object. We find this kind of work everywhere. It is divided into many professional branches and as a whole furthers the increasing aestheticization of reality. If we enumerate the different branches, we can see that they make up a large part of all social work. They include: design, stage sets, advertising, the production of musical atmospheres (acoustic furnishing), cosmetics, interior design – as well, of course, as the whole sphere of art proper. If we examine these areas in order to apply their accumulated knowledge fruitfully to aesthetic theory, it becomes apparent that this knowledge is in general implicit, tacit knowledge. This is explained in part by the fact that craft capacities are involved, which can scarcely be passed on by word but require the master's demonstration to the pupil. In part, however, the lack of explicit knowledge is also ideologically as a result of aesthetic theories. Although in practice something completely different is done, it is understood as giving certain things and materials certain properties. Occasionally, however, one finds an explicit knowledge that aesthetic work consists in the production of atmospheres.

Since knowledge about the production of atmospheres is very seldom explicit and in addition distorted by the subject–object dichotomy, I will go back to a classic example. I am referring to the theory of garden art, more exactly the English landscape garden or park, as it is presented in the five-volume work of Hirschfeld.[17] Here we find explicitly indicated how "scenes" of a certain quality of feeling can be produced through the choice of objects, colors, sounds, etc. It is interesting to note the closeness to the language of stage settings. By scenes, Hirschfeld means natural arrangements in which a certain atmosphere prevails such as serene, heroic, gently melancholic, or serious.

Hirschfeld presents, for instance, the gently melancholic scene in such a manner that it becomes clear how this atmosphere can be produced:

> The gently melancholy locality is formed by blocking off all vistas; through depths and depressions; through thick bushes and thickets, often already through mere groups of (closely planted) thickly leaved trees, whose tops are swayed by a hollow sound; through still or dully murmuring waters, whose view is hidden; through foliage of a dark or blackish green; through low hanging leaves and widespread shadow; through the absence of everything which could announce life and activity. In such a locality light only penetrates in order to protect the influence of darkness from a mournful or frightful aspect. Stillness and Isolation have their home here. A bird which flutters around in cheerless fashion, a wood pigeon which coos in the hollow top of a leafless oak, and a lost nightingale which laments its solitary sorrows – are sufficient to complete the scene.[18]

Hirschfeld indicates clearly enough the different elements through whose interaction the gently melancholy atmosphere is produced: seclusion and stillness; if there is water, it must be slow moving or even almost motionless; the locality must be shady, light only sparse in order to prevent a complete loss of mood; the colors dark – Hirschfeld speaks of a blackish green. Other parts of his book, which are more concerned with means, are even clearer. Thus, for instance, in the chapter on water:

> The darkness by contrast, which lies on ponds and other still waters, spreads melancholy and sadness. Deep, silent water, darkened by reeds and overhanging bushes, which is not brightened even by sunlight, is very suitable for benches dedicated to these feelings, for hermitages, for urns and monuments which sanctify the friendship of departed spirits.

Similarly, in the section on woodland he writes:

> If the wood consists of old trees reaching up to the clouds and of a thick and very dark foliage, then its character will be serious with a certain solemn dignity which calls forth a kind of respect. Feelings of peace possess the soul and involuntary cause it to be carried away by a calm contemplation and gentle amazement.[19]

The knowledge of the landscape gardener thus consists according to Hirschfeld in knowing by means of what elements the character of a locality is produced. Such elements are water, light, and shade, color, trees, hills, stones, and rocks, and finally also buildings. Hirschfeld thus recommends the placing of urns, monuments, or hermitages in the gently melancholic locality.

The question naturally poses itself as to what role these elements play in the production of the atmosphere as a whole. It is not sufficient to point out that the whole is more than the parts. With garden art we find ourselves in a certain way in reality itself. The same atmospheres, however, can also be produced through words or through paintings. The particular quality of a story, whether read or heard, lies in the fact that it not only communicates to us that a certain atmosphere prevailed somewhere else but that it conjures up this atmosphere itself. Similarly, paintings which depict a melancholy scene are not just signs for this scene but produce this scene itself. We could thus surmise that the components of a locality enumerated by Hirschfeld are not composed in just any fashion but that they conjure up an atmosphere.

Two aesthetic forms of production as different as that of the garden architect and the writer demonstrate a high degree of consciousness as regards the means by which particular atmospheres can be produced. A comprehensive investigation of the whole spectrum from stage designer to cosmetician would certainly throw new light on aesthetic objects, including works of art. Their "properties" would be understood as conditions of their atmospheric effect.

Conclusion

The new aesthetics is first of all what its name states, namely a general theory of perception. The concept of perception is liberated from its reduction to information processing, provision of data or (re)cognition of a situation.

Perception includes the affective impact of the observed, the "reality of images," corporeality. Perception is basically the manner in which one is bodily present for something or someone or one's bodily state in an environment. The primary "object" of perception is atmospheres. What is first and immediately perceived is neither sensations nor shapes or objects or their constellations, as gestalt psychology thought, but atmospheres, against whose background the analytic regard distinguishes such things as objects, forms, colors etc.

The new aesthetics is a response to the progressive aestheticization of reality. An aesthetics, which is a theory of art or of the work of art, is completely inadequate to this task. Moreover, since it is confined to a sphere separated from action and to educated elites, it hides the fact that aesthetics represents a real social power. There are aesthetic needs and an aesthetic supply. There is, of course, aesthetic pleasure but there is also aesthetic manipulation. To the aesthetics of the work of art we can now add with equal right the aesthetics of everyday life, the aesthetics of commodities, and a political aesthetics. General aesthetics has the task of making this broad range of aesthetic reality transparent and articulable.

Notes

1 Translated by David Roberts.
2 G. Böhme, *Für eine ökologische Naturästhetik*, Frankfurt/M., Suhrkamp, 1985.
3 Review by J. Früchtl, *Suddeutsche Zeitung*, 14 November 1989.
4 M. Seel, *Eine Ästhetik der Natur*, Frankfurt/M., Suhrkamp, 1991.
5 I refer to the cooperation with my brother Hartmut Böhme. See my interim report of 1989 "An Aesthetic Theory of Nature," *Thesis Eleven*, 32, 1992, pp. 90–102 and "Aussichten einer ästhetischen Theorie der Natur," in H.-G. Haben (ed.), *Entdecken verdecken*, Graz, Droschl, 1991.
6 N. Goodman, *Languages of Art*, Indianapolis, Bobbs-Merrill, 1968.
7 U. Eco, *Einführung in die Semiotik*, Munich, Frankfurt/M., 1972, p. 207.
8 E. Gombrich, "Maske und Gesicht," in E. Gombrich, A. Hochburg, and M. Plack, *Kunst, Wahrnehmung, Wirklichkeit*, 4th edn, Frankfurt/M., Suhrkamp, 1989, p. 10ff.
9 This denial is clearly evident in Eco's discussion of the advertising photo of a beer glass. "In reality," he writes, "when I see a glass of beer I perceive beer glass and coolness but I do not feel them. I feel rather visual stimuli, colors, spatial relations, the play of light" (ibid., p. 201). In this analysis the physiology of the senses gets in the way of the phenomenology of perception. The effect, in particular the effect of the advert, consists precisely in the fact that I really do feel "coolness" at the sight of the beer and that it is not simply a "period structure" which enables me to think of "ice-cold beer in a glass."
10 W. Benjamin, *Das Kunstwerk im Zeitalter seiner technischen Reproduzierbarkeit*, 11th edn, Frankfurt/M., Suhrkamp, 1979.
11 W. Benjamin, *Das Kunstwerk im Zeitalter seiner technischen Reproduzierbarkeit*, 1st vers. (my translation), Gesammelte Schriften, Frankfurt/M., Suhrkamp, 1991, pp. 12, 440.
12 L. Klages, *Vom kosmogonischen Eros*, 7th edn, Bonn, Bouvier, 1972, p. 93.
13 L. Klages, *Grundlegung der Wissenschaft vom Ausdruck*, 9th edn, Bonn, Bouvier, 1970.
14 Ibid., p. 72.

15 H. Schmitz, *System der Philosophie*, Bonn, Bouvier, 1964, vol. III, 2, p. 343.
16 Ibid.
17 C. C. L. Hirschfeld, *Theorie der Gartenkunst*, 5 vols., Leipzig, Weidmanns Erben und Reich, 1779–85.
18 Ibid., vol. I, p. 211.
19 Ibid., pp. 200 and 198ff.

2   Atmosphere as an aesthetic concept

"Atmosphere" is a colloquial term, yet despite or perhaps because of the ambiguity of its usage, it is helpful to return to it again and again. We speak of the tense atmosphere of a meeting, the light-hearted atmosphere of a day, the gloomy atmosphere of a vault; we refer to the atmosphere of a city, a restaurant, a landscape. The notion of atmosphere always concerns a spatial sense of ambience. An extraordinarily rich vocabulary may be used to describe it: cheerful, sublime, melancholy, stuffy, oppressive, tense, and uplifting. We also speak of the atmosphere of the 1920s, of a petit bourgeois atmosphere, of the atmosphere of power. The term itself, "atmosphere," derives from meteorology and, as a designation for an ambient quality, has a number of synonyms that likewise connote the airy, cloudy, or indefinite: these include climate, nimbus, aura, fluid; and perhaps emanation should be counted among them as well.

Between

As an aesthetic concept, atmosphere acquires definition through its relation to other concepts and through the constellations it creates in aesthetics. Atmosphere is the prototypical "between"-phenomenon. Accordingly, it is a difficult thing to grasp in words against the background of European ontology – Japanese philosophers have an easier time of it with expressions such as *ki* or *aidagara*. Atmosphere is something between the subject and the object; therefore, aesthetics of atmosphere must also mediate between the aesthetics of reception and the aesthetics of the product or of production. Such an aesthetics no longer maintains that artistic activity is consummated in the creation of a work and that this product is then available for reception, whether from a hermeneutical or a critical standpoint. An aesthetics of atmospheres pertains to artistic activity that consists in the production of particular receptions, or to the types of reception by viewers or consumers that play a role in the production of the "work" itself.

Atmospheres fill spaces; they emanate from things, constellations of things, and persons. The individual as a recipient can happen upon them, be assailed by them; we experience them, in other words, as something quasi-objective, whose existence we can also communicate with others. Yet they cannot be defined independently from the persons emotionally affected by them; they are subjective facts (H. Schmitz). Atmospheres can be produced consciously through objective arrangements, light, and music – here the art of the stage set is paradigmatic. But what they are, their character, must always be felt: by exposing oneself to them, one experiences the impression that they make. Atmospheres are in fact characteristic manifestations of the co-presence of subject and object.

Spatiality and presence

The aesthetics of atmospheres shifts attention away from the "what" something represents, to the "how" something is present. In this way, sensory perception as opposed to judgment is rehabilitated in aesthetics and the term "aesthetic" is restored to its original meaning, namely the theory of perception. In order to perceive something, that something must be there, it must be present; the subject, too, must be present, physically extant. From the perspective of the object, therefore, the atmosphere is the sphere of its perceptible presence. Only from the perspective of the subject is atmosphere perceived as the emotional response to the presence of something or someone. Aesthetics thus becomes the study of the relations between ambient qualities and states of mind, and its particular object consists in spaces and spatiality. Accordingly, it liberates things and works of art from the form in which their own reception was embedded and considers them in their ecstasies, i.e. with respect to the way in which they alter spaces by their presence. Where time had previously been dominant, what this aesthetics rehabilitates or discovers is above all spatiality – as when evening or night are studied as spatial phenomena or music as atmosphere. In contrast to the ubiquity of telecommunication, therefore, it focuses attention on locality and physical presence.

Performance and event

The aesthetics of atmospheres corresponds to a primary direction in the development of modern art. If in the visual arts we are dealing with paintings that represent nothing, and in literature with texts that have no meaning, then semiotics and hermeneutics cannot constitute the whole of aesthetics. Installation, performance, and happenings bring to light a dimension that always belonged to art, but was repressed in favor of form and meaning. The reproducibility of a work of art depends on the dissociation of its form from its specific concrete manifestation; it was here that Walter Benjamin identified the loss of aura. In ephemeral art, in the insistence on performance and event to the point of the denial of the work itself, artists attempt to reinvest their work with aura (D. Mersch).

Staging

The aesthetics of atmospheres is capable of addressing a broad spectrum of aesthetic work which, in traditional aesthetics, occupied marginal place or at most was labeled as "applied art," ranging from architecture and stage design to design and advertising. This is the area in which the desired transformation of art into life was actually accomplished by the avant-garde. Today there is no area of life, no product, no installation or collection that is not the explicit object of design. What was still a revolutionary act in art – the departure from the object – is here a method. For all the talk of design, at stake are not the things and their form. Rather, the focus is on scenes, life spaces, charisma. Here, atmosphere is the explicit object and the goal of aesthetic action. The aesthetics of atmosphere directs attention to what had always taken place in these areas of aesthetic work, though an ontology oriented to the thing had distorted it; the object and goal of aesthetic work is literally nothing; i.e. that which lies "between," the space. The architect may share facades and views with the painter, but what belongs to the architect is the shaping of space with confinement and expanse, direction, with lightness and heaviness. To be sure, the designer also gives objects form. But what matters is its radiance, its impressions, the suggestions of motion. Naturally, in advertising, information and representation are important too – but much more so the staging of products and their presentation as ingredients of a lifestyle.

Construction and criticism

Atmospheres are experienced as an emotional effect. For this reason, the art of producing them – above all in music, but also throughout the entire spectrum of aesthetic work, from the stage set to the orchestration of mass demonstrations, from the design of malls to the imposing architecture of court buildings – is at every moment also the exercise of power. In analyzing how atmospheres are produced, the aesthetics of atmospheres will hardly provide instruction for practitioners; rather the reverse – aesthetics must learn from practitioners. It will, however, afford a necessary critical potential, merely by saying that what one does has in most cases already happened. Today, aesthetics is no longer by any account the beautification of life or the appearance of reconciliation; rather, with the aestheticization of politics (Benjamin) and the staging of everyday life (Durth), it has itself become a political power and an economic factor.

3   The art of the stage set as a paradigm for an aesthetics of atmospheres

Making atmospheres

Our general theme, making atmospheres, has a provocative character. It sounds slightly perverse, even paradoxical. Making – does that not have to do with something tangible? With the world of concrete things and apparatuses? And atmosphere – is that not something airy, indefinite, something which is simply there and comes over us? How is one supposed to make atmospheres? Well, there is one sphere in which that has actually been going on for a long time: the art of the stage set. In it we have a paradigm which not only encourages us in our enterprise but endows the idea of making atmospheres with objective reality.

Atmosphere – a familiar yet extremely vague phenomenon

The term atmosphere has its origin in the meteorological field and refers to the earth's envelope of air which carries the weather. It is only since the eighteenth century that it has been used metaphorically, for moods which are "in the air," for the emotional tinge of a space. Today this expression is commonly used in all European languages, no longer does it seems artificial and is barely regarded as a metaphor. One speaks of the atmosphere of a conversation, a landscape, a house, the atmosphere of a festival, an evening, a season. The way in which we speak of atmospheres in these cases is highly differentiated – even in everyday speech. An atmosphere is tense, light-hearted or serious, oppressive or uplifting, cold or warm. We also speak of the atmosphere of the "petty bourgeoisie," the atmosphere of the 1920s, the atmosphere of poverty. To introduce some order into these examples, atmospheres can be divided into moods, phenomena of synesthesia, suggestions for motions, communicative, and social–conventional atmospheres. What matters is that, in speaking of atmospheres, we refer to their character. With this term character we already bring our understanding of atmospheres close to the sphere of physiognomics and theater. The character of an atmosphere is the way in which it communicates a feeling to us as participating subjects. A solemn atmosphere has the tendency to make my mood serious, a cold atmosphere causes me to shudder.

The scholarly use of the term atmosphere is relatively new. It began in the field of psychiatry, specifically in Hubert Tellenbach's book *Geschmack und Atmosphäre*[1] (Taste and Atmosphere). Here, atmosphere refers to something bordering on the olfactory – such as the climate of the homeland or the smell of the nest, that is, a sphere of familiarity which is perceptible in a bodily sensuous way. Since then, atmospheres have been researched in detail by phenomenology. Talk about atmospheres plays a part today in interior design, town planning, advertising, and all fields related to the art of the stage set – that is, the creation of backgrounds in radio, film, and television. In general, it can be said that atmospheres are involved wherever something is being staged, wherever design is a factor – and that now means: almost everywhere.

Now, this matter-of-fact way in which atmospheres are talked about and manipulated is extremely surprising, since the phenomenon of atmosphere is itself something extremely vague, indeterminate, intangible. The reason is primarily that atmospheres are totalities: atmospheres imbue everything, they tinge the whole of the world or a view, they bathe everything in a certain light, unify a diversity of impressions in a single emotive state. And yet one cannot actually speak of "the whole," still less of the whole of the world; speech is analytical and must confine itself to particulars. Moreover, atmospheres are something like the aesthetic quality of a scene or a view, the "something more" that Adorno refers to in somewhat oracular terms in order to distinguish a work of art from a mere "piece of work"; or they are "the Open" which, since Heidegger, has given us access to the space in which something appears. Seen in this way, atmospheres have something irrational about them, in a literal sense: something inexpressible. Finally, atmospheres are something entirely subjective: in order to say what they are or, better, to define their character, one must expose oneself to them, one must experience them in terms of one's own emotional state. Without the sentient subject, they are nothing.

And yet: the subject experiences them as something "out there," something which can come over us, into which we are drawn, which takes possession of us like an alien power. So, are atmospheres something objective after all? The truth is that atmospheres are a typical intermediate phenomenon, something between subject and object. That makes them, as such, intangible, and means that – at least in the European cultural area – they have no secure ontological status. But for that very reason it is rewarding to approach them from two sides, from the side of subjects and from the side of objects, from the side of reception aesthetics and from the side of production aesthetics.

Reception aesthetics and production aesthetics

The conception of atmospheres as a phenomenon has its origin in reception aesthetics. Atmospheres are apprehended as powers which affect the subject; they have the tendency to induce in the subject a characteristic mood. They come upon us from we know not where, as something nebulous, which in the eighteenth century might have been called a je ne sais quoi, they are experienced as something numinous – and therefore irrational.

The matter looks different if approached from the side of production aesthetics, which make it possible to gain rational access to this "intangible" entity. It is the art of the stage set which rids atmospheres of the odor of the irrational: here, it is a question of producing atmospheres. This whole undertaking would be meaningless if atmospheres were something purely subjective. For the stage set artist must relate them to a wider audience, which can experience the atmosphere generated on the stage in, by and large, the same way. It is, after all, the purpose of the stage set to provide the atmospheric background to the action, to attune the spectators to the theatrical performance and to provide the actors with a sounding board for what they present. The art of the stage set therefore demonstrates from the side of praxis that atmospheres are something quasi-objective. What does that mean?

Atmospheres, to be sure, are not things. They do not exist as entities which remain identical over time; nevertheless, even after a temporal interruption they can be recognized as the same, through their character. Moreover, although they are always perceived only in subjective experience – as a taste or a smell, for example, to return to Tellenbach – it is possible to communicate about them intersubjectively. We can discuss with one another what kind of atmosphere prevails in a room. This teaches us that there is an intersubjectivity which is not grounded in an identical object. We are accustomed, through the predominant scientific mode of thinking, to assume that intersubjectivity is grounded in objectivity, that detection of the presence and determinateness of something is independent of subjective perception and can be delegated to an apparatus. Contrary to this, however, the quasi-objectivity of atmospheres is demonstrated by the fact that we can communicate about them in language. Of course, this communication has its preconditions: an audience which is to experience a stage set in roughly the same way must have a certain homogeneity, that is to say, a certain mode of perception must have been instilled in it through cultural socialization.

Nevertheless, independently of the culture-relative character of atmospheres, their quasi-objective status is preserved. It manifests itself in the fact that atmospheres can be experienced as surprising, and, on occasions, in contrast to one's own mood. An example is when, in a cheerful mood, I enter a community in mourning: its atmosphere can transform my mood to the point of tears. For this, too, the stage set is a practical proof.

Phantastike techne

All the same, can one really make atmospheres? The term making refers to the manipulating of material conditions, of things, apparatus, sound, and light. But atmosphere itself is not a thing; it is rather a floating in-between, something between things and the perceiving subjects. The making of atmospheres is therefore confined to setting the conditions in which the atmosphere appears. We refer to these conditions as generators.

The true character of a making, which does not really consist in producing a thing, but in making possible the appearance of a phenomenon by establishing conditions, can be clarified by going back to Plato's theory of mimesis.

In the dialogue *Sophist*, Plato draws a distinction between two kinds of performing art, in order to unmask the mendacious art of the Sophists (Sophist, 235e3–236c7). There is a difference, he argues, between *eikastike techne* and *phantastike techne*. It is the latter which interests us here. In *eikastike techne*, mimesis consists in the strict imitation of a model. *Phantastike techne*, by contrast, allows itself to deviate from the model. It takes account of the viewpoint of the observer, and seeks to make manifest what it represents in such a way that the observer perceives it "correctly." Plato bases this distinction on the practice of the sculptors and architects of his time. For example, the head of a very tall statue was made relatively too large, so that it did not appear too small to the observer, or the horizontal edges of a temple were curved slightly upwards, so that they did not seem to droop to the observer.[2] This art of *phantastike* is perhaps not yet quite what we mean by the art of making atmospheres, but it already contains the decisive feature: that the artist does not see his actual goal in the production of an object or work of art, but in the imaginative idea the observer receives through the object. That is why this art is called *phantastike techne*. It relates to the subject's power of representation, to the imagination or *imaginatio*. We come close to what concerns us through the *skenographia* developed by the Greeks as early as the fourth century BC. In his Poetics 1449a18 Aristotle ascribes this to the tragedian Sophocles. The classical philologists believed that *skengraphia* already implied perspective painting, an invention frequently attributed to the Renaissance.[3] They claimed that the geometrical doctrine of proportion, in particular the theorem of radiation we find developed in the *Elements of Euclid*, was derived from *skenographia*. For in order to create spatial depth through painting, perspectival foreshortening of the objects represented – buildings, trees, people – is needed. In scenography, therefore, we have an art form which is now directed explicitly, in its concrete activity, toward the generation of imaginative representations in the subjects, here the audience. It does not want to shape objects, but rather to create phenomena. The manipulation of objects serves only to establish conditions in which these phenomena can emerge. But that is not achieved without the active contribution of the subject, the onlooker. It is interesting when Umberto Eco[4] claims precisely this for all pictorial representation: it does not copy the object, he asserts, but only creates the conditions of perception under which the idea of the object appears for the viewer of the image. That may be overstated, yet it is true for Impressionist painting, for example. That painting does not aim to copy an object or a landscape, but rather to awaken a particular impression, an experience in the onlooker. The most convincing proof of this is the technique of pointillism. The colors the painter wishes the onlooker to see are not located on the painted surface but "in space," or in the imagination of the onlooker.

Of course, the art of the stage set has by now advanced beyond pure scenography. Wagner's operas seem to have given particular impetus to this development, first because they demanded a fantastic ambiance in any case and, second, because they were intended to act especially on the feelings, not just the imagination.[5] But the breakthrough came only in the twentieth century, with the mastery of light and sound through electrical technology. Here, a stage art has now been developed which is no longer confined to the design and furnishing of the stage space but, on the one hand, causes the action on the stage to appear in a particular light and, on the other, creates an acoustic space which tunes the whole performance. At the same time, this has made it possible for the art of the stage set to leave the stage itself and spill over into the auditorium, or even into space itself. The spaces generated by light and sound are no longer something perceived at a distance, but something within which one is enclosed. This has also enabled the art of the stage set to expand into the general art of staging, which has applications, for example, in the decor of discotheques and the design of large-scale events such as open-air festivals, opening ceremonies of sports events, etc.[6]

The present dominance of light and sound design also enables us to discern in retrospect what the making of atmospheres consists of in the more object-related field. It becomes clear that what is at issue is not really visual spectacles – as was perhaps believed by practitioners of the old scenography – but the creation of "tuned" spaces, that is to say, atmospheres. The making, as long as it concerns a shaping and establishing of the geometrical space and its contents, cannot therefore relate to the concrete qualities possessed by the space and the things within it. Or, more precisely: it does not relate to the determinations of things, but to the way in which they radiate outwards into space, to their output as generators of atmospheres. Instead of properties, therefore, I speak of ekstases – that is, ways of stepping outside oneself. The difference between properties and ekstases can be clarified by the antithesis between convex and concave: a surface which, in relation to the body it encloses, is convex, is concave in relation to the surrounding space.

We are concerned, therefore, with ekstases, with the expressive forms of things. We are not accustomed to characterizing things in terms of their ekstases, although they are crucial to design, for example. In keeping with our ontological tradition, we characterize things in terms of their material and their form. For our present purpose, however, the thing-model of Jacob Böhme is far more appropriate. He conceives of things on the model of a musical instrument (Böhme, 1922).[7] In these terms, the body is something like the sounding board of a musical instrument, while its outward properties, which Böhme calls "signatures," are moods which articulate its expressive forms. And, finally, what is characteristic of things is their tone, their "odor" or emanation – that is to say, the way in which they express their essence.

Tone and emanation – in my terminology, ekstases – determine the atmosphere radiated by things. They are therefore the way in which things are felt present in space. This gives us a further definition of atmosphere: it is the felt presence of something or someone in space. For this the ancients had the beautiful expression parousia. Thus, for Aristotle, light is the parousia of fire (De anima, 18b).[8]

Conclusion

What I, harking back to Plato, called *phantastike techne*, would no doubt today be called design. We have oriented ourselves here by a prototypical area of design: stage design. But for our purpose it is important to modify the traditional understanding of design, according to which design amounted merely to shaping or configuring. That understanding is already prohibited by the extraordinary importance of light and sound, not only in the field of the stage set but also in advertising, marketing, town planning, interior design. One might speak of a practical, or better: a poetic phenomenology, because we are dealing here with the art of bringing something to appearance. A term used by the phenomenologist Hermann Schmitz is very apt here: he speaks of a "technology of impression" (*Eindruckstechnik*).[9] Admittedly, this term is used polemically, being applied to the generation of impressions for propaganda purposes in the Nazi period, or what Walter Benjamin called the "aestheticizing of political life" (Benjamin, p. 269).[10] Let us therefore speak more generally of the art of staging. On the one hand, then, we have preserved the connection to our paradigm of the art of the stage set and, on the other, we have included in this expression the purpose for which atmospheres are predominantly generated today: the stage set is itself a part of the staging of a drama or opera. The art of atmospheres, as far as it is used in the production of open-air festivals or in the build-up to large sporting events such as the Football World Cup or the Olympic Games, is their staging. The role of the generation of atmospheres in marketing is that of staging commodities. The commodities themselves are valued, in the aesthetic economy where they now serve only relatively little to satisfy basic needs, for their staging-value, that is, they are valued to the extent that they help individuals or groups to stage their own lifestyles. And, finally, in democracies, or more precisely media-democracies, in which politics is performed as if in a theater, the generation of atmospheres has the function of staging personalities or political events.

If we review these examples, it emerges that the attention which is now paid to atmospheres in aesthetic theory has its material background in the fact that staging has become a basic feature of our society: the staging of politics, of sporting events, of cities, of commodities, of personalities, of ourselves. The choice of the paradigm of the stage set for the art of generating atmospheres therefore mirrors the real theatricalization of our life. This is why the paradigm stage set can teach us so much, in theoretical terms, about the general question of the generation of atmospheres, and therefore about the art of staging.

But, in practical terms, too, there ought to be much to be learned from the great tradition of stage set design. That will happen indeed, but one should not expect that it will be possible to say very much about it. For the art of the stage set has been transmitted up to now, like traditional crafts, in master–pupil relationships, by collaboration and imitation. The guiding practical knowledge is tacit knowledge. It is all the more pleasing to find now and then, in the many books which exist on the subject of the stage set, something explicit about the craft. In conclusion, I will give an example of such knowledge from the praxis of the stage set. It is found in, of all places, a philosophical dissertation, Robert Kümmerlen's book *Zur Aesthetik bühnenräumlicher Prinzipien*.

Kümmerlen writes about the use of light on the stage. He argues, we should note, that an atmosphere is created on the stage with light. He then defines the effect of the light-atmosphere more precisely by saying that a characteristic mood is imparted by it to the performance. As examples, he mentions somber and charming moods – that is, moods with a synesthetic and a communicative character. Finally, he also recognizes the status of the "in-between existence" typical of atmospheres: "The lighting on its own generates a fluid between the individual structures of the performance." But now, let me give the quotation in full:

> The space to be contemplated is given its brightness by the lighting; stage performances are only made visible by light. The first function of lighting, the simple provision of light, creates, with the brightness, what might be called the atmosphere in which the space exists. The light-atmosphere, achieved in the most diverse ways, varies the space; through the lighting the performances take on a characteristic mood. The space creates an effect in its totality; the lights of the special representation produce a self-contained impression; the space stands in a unifying light. With the illumination of the whole scene a "unified character" is produced. A uniform mood emanates from the space; for example, the representation of space is subjected to a "muted" light. We find that three-dimensional objects "gleam" in a regular light; the space appears, for example, as "charming" or "somber." The lighting on its own generates a fluid between the individual structures of the performance. A specific mood is contained in the space represented through the ethereal effect of brightness.
>
> (Kümmerlen, p. 36)[11]

Notes

1 Hubert Tellenbach, *Geschmack und Atmosphäre*, Salzburg, Gebundene Ausgabe, 1968.
2 Carl Lamb, Ludwig Curtius, *Die Tempel von Paestum*, Leipzig, Insel-Verlag, 1944, p. 17.
3 Erich Frank, *Platon und die sogenannten Pythagoreer*, Darmstadt, WGB, 1962, p. 20.
4 Umberto Eco, *A Theory of Semiotics*, Bloomington, IN, Indiana University Press, 1976.

5 Ottmar Schuberth, *Das Bühnenbild. Geschichte, Gestalt, Technik*, Munich, Georg D. W. Callwey, 1955, p. 86, 95.

6 Ralph Larmann, *Stage Design*, Cologne, Ppv Medien GmbH, 2010; Tony Davis, *Stage Design*, Ludwigsburg, Avedition, 2001.

7 Jacob Böhme, "De signature rerum oder von der Geburt und Bezeichnung aller Wesen," in Jacob Böhmes sämtliche Werke (ed. K. W. Schiebler), vol. I., Leipzig, 1922.

8 Aristotle, *De Anima: On the Soul*, trans. Mark Shiffman, Newburyport, MA, Focus Publishing/R. Pullins, 2011.

9 Hermann Schmitz, *Adolf Hitler in der Geschichte*, Bonn, Bouvier, 1999.

10 Walter Benjamin, "The Work of Art in the Age of Its Technical Reproducibility" [translated by Edmund Jephcott], in W. Benjamin,*Selected Writings*, vol. 4, Cambridge, MA and London, Harvard University Press, 2003.

11 Robert Kümmerlen, *Zur Aesthetik bühnenräumlicher Prinzipien*, Ludwigsburg, Erscheinungsjahr, 1929.

4   Kant's aesthetics: a new perspective[1]

Kant's negative aesthetics

I would like to make the suggestion that Kant's *Critique of Judgment* be read "laterally." This suggestion is in reaction to the frustrating sight offered by current Kant research: here the attempt is made over and over again to comprehend Kant's text, that is, to follow its train of argument and justify its systematic claims, or to rectify its inadequacies. It arises, on the other hand, from the experience of the interminable struggle students have with the reading of what is, in itself, a rich and fine piece of writing. The appropriateness of reading the *Critique of Judgment* in another way, must, of course, prove itself by its fruitfulness. We can, however, legitimate it in advance by noting that Kant's initial readers, his contemporaries, themselves read him "laterally." Schiller for instance, for whom the concept of beauty was opened up by §59 ("Beauty as the symbol of morality"), that is, by the concept of freedom. An explicit recommendation to "read laterally" is given by Goethe:

> As I sought, if not to penetrate, then, nevertheless, to make most use of Kant's doctrine, it seemed to me, at times, as if the good man was pro-ceeding in a mischievously ironic fashion: one moment he seemed at pains to restrict knowledge to the narrowest of confines, and in the next he was gesturing, with a wink, beyond the limits he had himself just drawn ... Then, after he had sufficiently cornered us, indeed brought us to the brink of despair, he opts for the most liberal of statements and leaves it up to us to decide what use we wish to make of this freedom, which he concedes, in some degree.[2]

"If not penetrate, then, nevertheless, to make use of": this key phrase indicates that Goethe, in his free use of the *Critique of Judgment*, took Kant far more seriously than can conservative and systematizing Kant research, namely as a direct partner in the engagement with actual problems in the field. We want to follow his example.

My interpretative hypothesis is the conjecture that Kant, through his logical or transcendental–logical approach, can express what he wants to say in the

Aesthetics only indirectly and with distortion. As a result, Kant's most important and fruitful assertions are often found in footnotes and excurses or even in places which, strictly speaking, fall outside his system. This observation, incidentally, holds for issues other than the *Aesthetics*: Kant was a highly sensitive and engaged human being, but one whose rational discipline of thought and way of life meant there was much which he could not permit himself. His most important observations and statements are, for the reader, often hidden in his work.

Kant's aesthetics is not written as aesthetics but as a theory of judgment. It therefore does not treat, or at least not directly, the experience of the beautiful, but contains an analysis of judgments of the type "x is beautiful," judgments of taste. The text is entirely shaped, that is, in construction and execution, according to a schema that Kant developed elsewhere and to other purposes, namely the schema of the judgment – or categories – table, and the schema divided up into various analytics, dialectics, deductions, and methodology. The violence done to the thematic content of the text by this schema is evident in the fact that the subject matter is always breaking through the schema, and demonstrates itself finally in the historical effects of Kant's text. For it is precisely through its reception of this text that German Idealism shatters the Kantian System.

I would like to describe Kant's *Critique of Judgment* as a "negative aesthetics." In doing so I am playing on the thought of a negative theology as well as upon titles such as "Negative Dialectics" and "Negative Anthropology." Kant allows the beautiful to become visible by stating what it is not. This shows itself most clearly in his "elucidations of the beautiful" which follow the four-way schema of the table of judgment. An object is beautiful when it is the object of pleasure:

1　apart from any interest (§5)
2　apart from a concept (§9)
3　apart from the representation of an end (§17).[3]

The only positive determination is to be found in the fourth "Moment": "The beautiful is that which, apart from a concept, is cognized as object of a *necessary delight*" (§22). Insights into the essence of the beautiful are obtained through all these negative determinations, if indirectly. This should be made clear immediately by one of the most important of the negative characterizations of beauty, namely Kant's assertion *that beauty is not a predicate*. The judgment of taste – judgment in the form "x is beautiful" – does not say, precisely speaking, that x has the characteristic of being beautiful. What is involved is not properly a predication. This improper predication is the well-known figure of the "as if." This figure permeates the whole *Critique of Judgment*, is found in the analysis of the beautiful, the sublime, and equally determines the whole teleology.[4] Of the beautiful, in §6 Kant says: "He (the judge) will speak of the beautiful as if beauty were a quality of the object and the judgment logical."

In fact, this is how one speaks: the dress is beautiful, this tulip is beautiful. This form of speech suggests that the term "beautiful" is a predicate that applies to the subject "dress," "tulip." Kant's critique, the critique of judgment, consists, precisely, in breaking through this appearance. But what, one might ask, does this "beautiful" refer to, if not the object's dress or tulip, which comprise the logical subject of the sentence? In the case of the sublime Kant gives an unambiguous and decisive answer to this question. What is actually sublime are not those objects, which we improperly predicate with sublimity, but we ourselves, the subjects who experience them as sublime.

> This makes it evident that true sublimity must be sought only in the mind of the judging Subject, and not in the Object of nature that occasions this attitude by the estimate formed of it. Who would apply the term "sublime" even to shapeless mountain masses towering one above the other in wild disorder, with their pyramids of ice, or to the dark tempestuous ocean, or such like things? But in the contemplation of them, without any regard to their form, the mind ... feels elevated in its own estimate of itself.[5]

When it comes to the beautiful Kant is not so unambiguous and so decisive. But it nevertheless becomes clear that the basis of the judgment "x is beautiful" is to be sought, not in x, but in the judging and aesthetic subject. If the analogy to the judgment "x is sublime" were to be taken quite strictly, then the true meaning of "x is beautiful" would have to be "I feel myself beautiful (in the face of x)." Kant avoids this strict analogy, because he places a value on the idea that the feeling of pleasure in the face of the beautiful is not itself the foundation of the judgment "x is beautiful," but only a necessary consequence of this judgment. The judgment "x is beautiful" is nonetheless a judgment on the state of the subject (in the face of the beautiful), namely a judgment on the concerted play of the powers of imagination. "The determining ground of judgment" is not any "concept of the object," but the "mental State." Here is the complete quotation:

> If, then, the determining ground of the judgment as to this universal communicability of the representation is to be merely subjective, that is to say, is to be conceived independently of any concept of the object, it can be nothing else than the mental state that presents itself in the mutual relation of the powers of representation so far as they refer a given representation to cognition in general?[6]

The situation, which Kant sees confronting him in the analysis of the beautiful, is that of *a relation* between subject and object. That he is aware of this situation will become quite clear if I repeat more extensively the section from §6 already referred to:

> He will speak of the beautiful as if beauty were a quality of the object and judgment logical (forming a cognition of the object by concepts of

it); although it is only aesthetic, and contains *merely a reference* of the representation of the object to the Subject. [my emphasis]

Kant remains sufficiently under the spell of traditional metaphysics, namely the ontology of substance, that he cannot conceive relation as such, but must attribute the content of the relation to its component parts. The relation must always possess a *fundamentum in re*. As he cannot locate the basis of the relationship in the object, he seeks it in the perceiving subject. We must ask ourselves whether this decision was an obligatory one for Kant and furthermore, whether it is adequate to the situation.

After all, the most obvious treatment of the predicate "beautiful" would have been to treat it analogously to secondary qualities. In the case of predicates like "warm" or "blue" it had been clear to the philosophy of the modern era, at least since Locke, that what was being dealt with here are qualities attributable, not to objects themselves, but only in relation to a knowing subject, that is, as appearances. As the object in any case remains object in the appearance, the secondary qualities are also objective. Why does Kant treat the predicate "beautiful" differently to the predicates "blue" or "warm"? I suspect that the reason is to be found, not in the logical analysis of the judgments of taste, but in the experience of beauty itself.

A further possibility, which would have been open to Kant, lies in the assertion, which I somewhat prematurely put forward as the actual result of this analysis, namely that beauty is not a predicate. Kant does not actually express himself this way, but merely says, with his formula of the "as if," that what is being dealt with is not properly a predication, and, in the case of the sublime, states that the predicate "sublime" actually applies to the cognizing subject. But was it not also possible for Kant to argue that beauty is not a predicate at all?

This possibility seems to me to be given in principle, if one brings into consideration Kant's treatment of the ontological proof for the existence of God. Here Kant takes as point of departure a tradition, which he analyzed as expressing the view that existence is a predicate of God. According to Kant's analysis the ontological proof takes the following form:

God is the most perfect Being.
Assuming he were lacking in existence, then a Being that possessed all the qualities of God, yet also existed, would be more perfect.
Therefore existence must also be among God's predicates.

Kant says that the flaw of this argument lies in its treatment of existence as a predicate. Being is not a predicate. If Kant thought along these lines, then could he not equally have maintained that "being beautiful" (*Schön-sein*) is not a predicate? If he could not draw this conclusion, this is clearly due to thinking, in line with tradition, of being (in the sense of existence) as purely without qualities, that is, he made no distinction, as we might express it today, between modes of being. For him the existence of something is not a predicate, but rather a modality of judgment. To say "x exists," means, according to the analysis of the doctrine of postulates in the *Critique of Pure Reason*, that the object, thought of as x, is given to sensation. We can see at this point that it would certainly be possible to analyze beauty in analogy to being.

That Kant actually comes close to such an analogy will now be demonstrated. The improper predication of beauty in the judgment of taste, according to Kant, refers fundamentally to a singular object, more precisely: to a "this there." In the citation of the examples, I have attempted to formulate this clearly in each case. "This tulip is beautiful" is not a judgment of taste in the sense of "all tulips are beautiful," but only in the sense of "this tulip there is beautiful." To illustrate this, I cite a quotation from §33, which clearly emphasizes this characteristic of the judgment of taste.

> As a matter of fact, the judgment of taste is invariably laid down as a singular judgment upon the object. The understanding can, from the comparison of the object, in point of delight, with the judgment of others, form a universal judgment, e.g.: "All tulips are beautiful." But that judgment is then not one of taste, but is a logical judgment, which converts the reference of an object to our taste into a predicate belonging to things of a certain kind. But it is only the judgment whereby I regard an individual given tulip as beautiful, i.e., regard my delight in it as of universal validity, that is a judgment of taste.[7]

But perhaps this quotation will be found still insufficiently clear, because it merely distinguishes, from a logical standpoint, that is, not from a transcendental standpoint, specific from general judgments. From this standpoint, one might believe that the judgment of taste, because it does not refer to tulips in general, instead refers to a single, that is, particular given tulip. But examined more closely, one finds one can only understand the peculiarity of the judgment of taste transcendentally, that is, through taking into account the mode of givenness: the judgment "x is beautiful" can only be made at all, if and insofar as x is given. The negative determination of the judgment of taste, namely that it is "apart from a concept" (§9), means also that this judgment does not refer to the x, insofar as it can be thought. It refers namely to x in its givenness. As a proof I quote the sections of §8, which refer to this peculiarity of the judgment of taste. There Kant states:

> In their logical quantity, all judgments of taste are *singular* judgments. For, since I must present the object immediately to my feeling of pleasure or displeasure, and that too without the aid of concepts, such judgments cannot have the quantity of judgments with objective-general validity.

"I must present the object immediately to my feeling of pleasure or displeasure": that is the decisive point. Shortly after this it continues "We want to get a look at the object with our own eyes."[8] The consequence is: the judgment of taste "x is beautiful" can only be made in the presence, or on the basis of the experienced presence of x. This judgment cannot communicate that the object possesses this or that characteristic, by only that one responds to it in a characteristic way in its presence. This characteristic response is indicated by the expression "beautiful." Beauty is therefore an experience of presence, more precisely: the experience of a co-presence. What is articulated in the judgment of the form "x is beautiful" is namely, that I, as a cognizing subject in the presence of x, "find" myself in a characteristic state. The state of mind in which one finds oneself – Kant repeatedly talks here about "moods" (*Stimmungen*) or "states of mind" (*Gestimmtheit des Gemüts*) – is namely itself a form of presence.

With this formulation, however, we have already stepped outside Kantian thought. For with it, we have attributed beauty as a predicate neither to the object nor the subject. Beauty is not a predicate, but the characteristic of a co-presence. Elsewhere I have described co-presences, distinguishable by characteristics as atmospheres.[9] Beauty, accordingly, is an atmosphere, a shared reality of subject and object.

Beautiful form: Kant's examples

Kant's aesthetic is carried out in the form of an analysis of the judgments of taste. A judgment of taste had the form "x is beautiful." What foundation, it might be asked, did Kant have for his analysis? This question really becomes clear only when the situation of the *Critique of Judgment* is compared to that of the *Critique of Pure Reason*. There the analysis of knowledge, that is, the analysis of objective judgments, was dealt with. The foundation for this is clear. It consists of the propositions of science and mathematics: for the question regarding the conditions of the possibility of experience, Kant presupposed the facts of science. Is there anything corresponding to this for the *Critique of Judgment*? What might that have been? There can be no talk of a canon of aesthetic judgments in the eighteenth century, of the kind recognized in mathematics or Newtonian science. What else would be conceivable as the foundation for a logical and transcendental–logical analysis of the judgments of taste? He could have presupposed, if not a canon of valid judgments of taste, then at least what such a judgment is. But he clearly does not do this. Rather, on the basis of his analysis, Kant defines a judgment of taste through its differentiation from an objective judgment. Furthermore, he could have investigated the use of the predicate "beautiful." But, if Kant had not also felt it legitimate to correct this language usage, this procedure would have led to merely empirical results. He also could have taken a critical look at the thoroughly developed aesthetic discourse of the eighteenth century, and continuing and distancing himself from this discourse, have developed his own aesthetics. Kant undoubtedly did occupy himself with this discourse and read the works of Diderot, Shaftesbury, Burke, Hutchinson, Winckelmann, and

Lessing. But – with the exception of Burke – there is no direct debate with these works.[10]

Finally, there remains the engagement with the phenomenon of beauty itself, or an analysis of the experience of the beautiful. In fact, I wish to maintain that Kant's text lives from the experience of the beautiful, and that, for this reason, its representation from the perspective of the *judgment* of the beautiful becomes a distortion and can only indirectly make visible what Kant actually has to say. If that is true, then his examples hold special weight for the understanding of the work.

Let us, through a few early sections of the text, once more call to mind the way in which Kant proceeds. He names Book One "Analytic of the Beautiful." But this analytic of the beautiful consists of running through the different logical moments of the judgment of taste. Here too, already present in the title, we have the enigmatic oscillation between the analysis of the phenomenon of beauty and its judgment. The first sentence reads:

> If we wish to discern whether anything is beautiful or not, we do not refer the representation of it to the object by means of understanding with a view to cognition, but by means of the imagination (acting perhaps in conjunction with understanding) we refer the representation to the Subject and its feeling of pleasure and displeasure.[11]

This sentence is not actually undertaking an analysis, but rather already presenting the result. Yet it is nevertheless revealing for our question, because it shows how Kant proceeds, namely through a kind of phenomenology of the judging experience: he observes what happens, when we decide (not distinguish), whether something is beautiful or not.

But this procedure, entirely appropriate for a transcendental formulation of the question, does not remain the sole method. In addition, he employs a distinction between the expressions "pleasant," "good," "beautiful." Here he proceeds normatively, that is, he establishes when the predicate beautiful is to be used in distinction to pleasant and good. He does this, admittedly, in descriptive language. Thus it is characteristic of judgments of something as beautiful, rather than as pleasant, that the pleasure is grounded exclusively in the judgment of the object, and not in its real existence.[12] The judgment of something as good, as opposed to the judgment of something as beautiful, presupposes a concept of the thing, in order namely to be able to judge it with regard to how appropriate it is to the concept. This normative–descriptive procedure, if it were made explicit, could involve the assertion that there are different types of judgment or forms of assessment. The task would then be to demonstrate this through examples. It is also from this point of view that Kant's examples acquire special interest.

It can be observed that Kant normatively regulates the use of the predicate "beautiful." Thus he safeguards his assertion, that the judgment of taste makes a claim to general validity, by determining the use of the predicate beautiful as illegitimate, where this is not the case: "For if it merely pleases *him*, he must not call it *beautiful*."[13] This normative procedure becomes even clearer, when he criticizes the usual, or until then usual, practices of applying the expression beautiful. Thus for example, one cannot, as is commonly the case, use "geometrically regular figures, a circle, a square, a cube and the like" as show-piece examples of beauty, and furthermore it is just as unjustified to attribute the predicate beauty to the changing forms of a hearth fire or a running stream.[14] This method justifies itself only when one can point to characteristic differences in the respective examples, and then build these up into normative language rules.

Let us now examine Kant's examples in the aesthetics itself. These must provide information as to what Kant actually envisaged with his analysis of the beautiful and what experience he underwent in this area. We will concentrate our attention on the examples of the beautiful, given that the sublime in Kant has already received extensive acknowledgment in recent times.[15]

If one observes the examples of beauty, it immediately stands out, as has already been noted many times,[16] that the references to works of art are extremely scant and rather incidental. Kant's knowledge of art appears to have been very limited and indirect. He mentions Muron's cow and the Doryphorus of Polyclitus, two insignificant poems, and music only in the vague form of "human song" or "concert." This "lack," however, strikes us only retrospectively, from the viewpoint of an aesthetics that has become in every respect an aesthetics of the artwork. Seen in this way, Kant's aesthetics is primarily an aesthetics of nature and only secondarily an aesthetics of art, insofar as art, in order to be beautiful, must resemble nature. If one wishes to regard Kant's aesthetics as an aesthetics of nature, a position fully justified on the basis of the examples, it becomes apparent, that alongside the group of examples drawn from nature, there is another equally powerful group derived from the area of what we might today call design. In the examples drawn from nature, what dominates is the "flower," only seldom specified as tulip or rose. In addition are to be found numerous birds and the crustaceans of the sea, then feathers, crystals, and among particular animals beside the nightingale there appears only the horse. We might find it strange in this list, that leaf and tree types and animal varieties are not investigated. Landscapes are markedly absent. Only once is there mention of a forest clearing: "if in a forest I light upon a plot of grass, round which trees stand in a circle." Beside this group of examples from nature, of particular importance is what we have called design. Here there is much discussion of dress, then of buildings of all kinds, where the talk is of the decoration of rooms, of tasteful table implements, furniture, carpets, and finally nature as design, namely the flowers with which one decorates the front garden, and ornamental gardens.[17]

Precisely the last observation is important: namely that the examples of the beautiful things of nature are also for the most part mentioned as components of design: foliage as ornament, the song of the nightingale as an arrangement for the gratification of country-house guests, flowers as decoration and adornment of the front garden.[18] What we call "wild nature" is scarcely mentioned by Kant, in any case not under the heading "beautiful," but rather much more frequently under the heading "sublime." "Free beauties of nature" (*freie Naturschönheiten*) are not beauties found in untamed nature (*in der freien Natur*), but those which are non-purposeful and non-conceptual, and this includes the designs of crustaceans just as much as the designs *à la grecque* on carpets.[19]

Equally the few examples drawn from or of art also appear in a constellation which we would generally designate today as design. Thus one of the two poems is an "academic poem," music appears as background music (*Tafelmusik*) and landscape gardening as a kind of painting.[20] Admittedly one must remember here that a clear distinction between art and handicraft does not exist, nor, as yet, a clear concept of autonomous art. Furthermore, Kant distinguishes in §44 between the "pleasant" arts (*angenehme Künste*) and the "fine" arts (*schöne Künste*). Yet it does not seem to follow from this distinction, that the product of the "pleasant" arts is not beautiful, but merely "pleasant." Freedom from purpose, which belongs to the concept of the beautiful, can be entirely upheld, as we shall see when we consider §41, when the beautiful serves the purpose of conviviality. Thus musical entertainment and table decor are rather examples of the beautiful and not of the pleasant.

On the whole, the impression is reinforced that the examples, from which Kant constructs his concept of the beautiful, belong neither to untamed nature nor to autonomous art, but rather much more to the sphere of the life world, which is beautified through decor, ornaments, arrangements, "finery." If one might say that Kant's experiences of great art were extremely limited and that the experience of wild nature was also one he hardly sought – even the examples of the sublime he derived largely from travel literature – Kant was by no means aesthetically insensitive. On the contrary in matters of taste he was obviously extremely sensitive – and an expert – but "taste" for him referred primarily to the components of cultivated social intercourse. "Taste is a requirement of social intercourse; and social intercourse the nourishment of taste."[21]

If one searches for aesthetic interests in Kant's biography, one might be dissuaded by the story of the severity of his pietistic youth and the austerity of his housekeeping in old age. These stories, however, are also stylizations and belong to the image of Kant as the stern moralist. Another stylization stands in contrast to them, namely that of the "elegant doctor."[22] The stories which present Kant to us as a sociable person, friend of conversation, games and fashion, belong rather to the middle of his life. They become of increasing significance and reconcilable with the image of Kant as moralist, if we recall that for Kant cultivation represents the necessary preliminary stage to becoming moral, and that sociability is considered almost as a moral duty. Let us examine one of the most important testimonies to Kant's aesthetic practice. This is from a section of Borowski's biography.

More than 40 years ago K. had impressed upon himself and, when he had the opportunity, upon us, his audience of the time, that a person's attire must never be found completely out of fashion; it is, he would add, one's absolute duty never to present to the world a distasteful, or even a curious, sight. He made it a closely observed maxim that the choice of color in dress and waistcoat should be directed in accordance with the example set by flowers. Nature, he would say, brings forth nothing which does not please the eye; the colors which she places side by side always suit one another. Thus, for example, a yellow waistcoat matches a brown overcoat; this is indicated to us by the *auriculae*. K. always dressed with refinement and respectability. Later on he particularly loved muted colors. For a while he could be observed in coats trimmed at the seam with little gold braid. He uprightly carried a rapier as long as businessmen wore them; but was more than happy, when this custom came to an end, to set it aside, as a rather cumbersome and superfluous accessory. Only his hat, as far as I observed, was never submitted to the law of fashion.[23]

Kant is a person of the rococo. His taste was formed by the furniture of Louis XV and Louis XVI, by "ornamented interiors," by the stucco of the late baroque and rococo, the garlands of flowers, tendrils, the sequences of mussels, it was formed by the shawls and folds of ladies clothing, the delicate and elegant form of silhouettes, by conviviality and conversation, meant to be playful and committed to cheerfulness. This entire world is an heir to courtly forms of life. In the bourgeois world it promotes the stylization of life and the concealment of seriousness, diligence, and business. In contrast to this disciplined world, the beautiful, as Schiller later observed, is the anticipation of the realm of freedom.[24] The "beautification" of life promotes playfulness, the enlivening of the imagination, the stimulation of the pleasure of life.

Kant's understanding of beauty concurred with that of William Hogarth, to whom his attention must have been drawn, at the latest, by Hamann in 1758, after his journeys to England. He mentions "Hogarth's burin" in his work *Observations on the Feeling of the Beautiful and the Sublime* (1763/4, A18). In *The Analysis of Beauty* Hogarth had characterized waving lines as "lines of beauty" and serpentine lines as "lines of grace."[25] He locates the origin of the experience of beauty in a stimulation of the mind and therefore calls for a "constant rule of composition in painting, to avoid regularity."[26] It is characteristic that Hogarth demonstrates his lines with examples which are, like Kant's drawn from design and fashion, thus, for instance, the lines of beauty in the curved chair legs of his time. The stimulating indeterminacy of form characteristic of beauty Hogarth calls intricacy.

Intricacy in form, therefore, I shall define to be that peculiarity in the lines, which compose it, that *leads the eye a wanton kind of chance*, and from the pleasure that gives the mind, entitles it to the name of beautiful.[27]

The beautiful is a matter of form but not of any form, rather of the form which is indeterminate, a *"je ne sais quoi,"*[28] a suggestion, a myriad potentiality. It is a rococo line which Kant envisages: light, playful, undecided, averse to repetition, one moment making a distinct promise and in the next moment denying it, the line of pure suggestion (*Anmutung*).

Communicability and sociability

One of the most important characteristics of the beautiful, according to Kant, is its communicability. This communicability has been brought into connection, rightly, with the proximity of the judgment of taste to the cognitive faculty and its claim to universality: for Kant it is indeed possible to dispute taste. Aesthetics subsequently developed as the foundation and instrument of art criticism, and in Habermas it ultimately becomes the discourse on the appropriateness of expression. But when one looks back from the prattle of catalogues and feuilletons, to what, according to Kant, can be said about aesthetics, this proves to be, strictly speaking, very little, almost nothing. It exhausts itself namely in the sentence "x is beautiful." But does Kant really mean this sentence, when he is speaking of the communicability of the beautiful and the ability to impart the judgment of taste? What is it that is imparted and in what way? As the beautiful is non-conceptual, it is indeed possible to pronounce the sentence "x is beautiful" but nothing further can be said about this x with regard to beauty. As we have already seen, the sentence "x is beautiful" is not, strictly speaking, a statement about x, but a communication of my feeling in the face of x. Indeed, when one studies the places in which Kant treats "communicability" in the *Critique of Judgment* more carefully, this reveals itself as the communication of a state of mind and not as a statement about an object. Thus the title of the decisive §39 reads "The communicability of a sensation." It says that the "one who judges with taste... can impute the subjective finality (*Zweckmässigkeit*), i.e., his delight in the object, to everyone else."[29] Shortly after Kant then specifies the situation of aesthetic communication: "the representation communicates itself not as thought, but as an internal feeling of a final state of the mind (*eines zweckmässigen Zustandes des Gemüts*)."[30] This communicability is therefore obviously not a discussion of the beautiful, but rather that co-participation in that state of mind, experienced in the face of the beautiful. This, admittedly, is not a matter for everyone, but strictly speaking belongs to the capacity of the genius, hence of the artist. He is in the position, "to hit upon the *expression* for them (i.e. the aesthetic ideas) – the expression by means of which the subjective mental condition induced by the ideas as the concomitant of a concept may be communicated to others."[31]

With this phrase Kant attributes a role to art in his aesthetics, which is not exhausted through being a second field of the beautiful, alongside nature. Art fulfils, in a fundamental sense, the call for the communicability of the beautiful. Because it deals with the communication of a feeling, the respective form of communication must be constituted quite differently from that of the statement, that is, the model of rational discourse. This other model, which deals with the communication of feelings and not the discussion about feelings, with the conveyance of feelings and with participation, naturally appears in its purest instance, where thoughts are entirely absent from the communication. Thus it is, from the perspective of communicability that art steps into the foreground, which has been dismissed by the disciplined thinker Kant and relegated to a minor position in reason's hierarchy of the arts: the art of music. Kant writes:

> Its charm, which admits of such universal communication, appears to rest on the following facts. Every expression in language has an associated tone suited to its sense. This tone indicates, more or less, a mode in which the speaker is affected, and in turn evokes it in the hearer also, in whom conversely it then also excites the idea which in language is expressed with such a tone. Further, just as modulation is, as it were, a universal language of sensations intelligible to every man, so the art of tone wields the full force of this language wholly on its own account, namely, as a language of the affections, and in this way, according to the law of association, universally communicates the aesthetic ideas that are naturally combined therewith.[32]

Kant is using here, as a base, a model of communication, first clearly formulated by Jakob Böhme.[33] This deals with a language of feelings, with expression and with an understanding that consists of an inner empathetic grasping of what the expression is an expression of.

But not everyone is a genius. Not everyone can create artworks which permit others to participate in his or her state of mind. Nonetheless the latter is expected from everyone, or, strictly speaking, from everyone with taste. Here it becomes important that taste as a capacity of judgment is, at the same time, a capacity to select. The ability to assess things as beautiful makes it possible for me to select beautiful things. Therefore, whoever cannot communicate his feeling through the creation of artworks, is nevertheless in the position, as a person of taste, to allow others to participate in his pleasure in objects, through his acquisition of them. As in the case of the artist, the communication here is also concrete. The person of taste, the "refined human being," as Kant expresses it, beautifies his surrounding, he surrounds himself with beautiful things and thereby permits others to share in his pleasure at the world. From here the "empirical interest in the beautiful" (§41) can now be explained. Kant names as interest that pleasure "which we connect with the representation of the real *existence* of an object."[34] Kant repeatedly accentuated that pleasure in the beautiful is always independent of interest, that is, it refers only to the "how," not the "that" of an object. The beautiful is free of purposiveness, and there is therefore, according to Kant, no desire for the object, or its existence. Now, however, it is revealed that there is a function of beauty after all, that one desires that it exists in order to fulfill that function, and that one rejoices in its existence: all this involves the fact that the beautiful allows the communication of one's feelings. The interest in the beautiful is therefore only indirect. One desires the beautiful, insofar as one desires society. Kant goes so far as to invert this thought. In the *Critique of Judgment* he writes "a man abandoned on a desert island would not adorn either himself or his hut, nor would he look for flowers, and still less plant them, with the object of providing himself with personal adornments. Only in society does it occur to him."[35] The lectures on logic, in Philippi's transcript, repeat, with greater clarity, if more crudely, the same thought:

> When I am alone, I am happier in the forest: in company the garden is preferable. Beautiful form seems to exist only for company. Why is it that taste disappears in isolation? In isolation we look only to our private pleasure. This need not follow judgments informed by the laws of taste. There (in company) we must direct ourselves according to the general pleasure, little as it may excite or please us. If we are alone, we never attend to the beautiful.[36]

Kant establishes, in his way, a relationship between the beautiful and the social, which Burke had already noticed. Burke had assigned society to beauty, but self-preservation to the sublime.[37] Beauty was thereby conceived as erotic, that is, as something which brought about attraction and proximity between individuals. In Kant also, beauty constitutes sociability, but indirectly, as a vehicle which renders feelings communicable. And the sociability which is established here is not the coarse society of sexual beings, but the higher one of cultivated beings, of people of taste. The important connection between the communicability of the judgments of taste and the empirical interest in the beautiful is thereby established.

The universality of judgments of taste makes them communicable in principle. What can be communicated is not a logical form or an objective statement, but a feeling. One can, even if not an artist, communicate one's feeling of pleasure in the beautiful, by acquiring it. By surrounding oneself with beautiful things, one cultivates the society of people of taste. All this belongs to the constellation of life for the cultivated bourgeoisie in the rococo. The beautiful here is neither the "intimation of the terrible" (Rilke), nor the object of erotic desire. The beautiful is neither something divine nor a power. It serves the cultivation of life. This is not trivial, for in this way moralization, and thereby true society, is prepared for. "Taste would then reveal a transition on the part of our critical faculty from the enjoyment of sense to the moral feeling."[38]

Education

This potential transition from a cultivated life world to a moral existence was extremely important for Kant. In his *Anthropology* he discusses it in two places, namely in §14, "On permissible moral semblance," and in §69, "Taste includes a tendency to promote morality externally." The logical and transcendental analysis of the judgment of taste pushes this perspective into the background. The empirical interest in the beautiful is, as is stated in the *Critique of Judgment* "of no importance for us here."[39] Nevertheless, Schiller perceived it as of the utmost importance and responded to Kant with his *Letters on the Aesthetic Education of Humanity*. And Schiller was right in doing so. For in fact the concept of education (*Bildung*) runs throughout the whole Kantian text.

Education plays a decisive role at three points in the Kantian aesthetics. First, for the universality of the judgments of taste; second, for the intellectual interest in the beautiful; and, third, for the receptivity to the sublime. In the *Critique of Judgment* Kant does not use the concept of education, but rather that of culture or civilization, without making a clear distinction between these two terms. In his *Pedagogy* and in his *Anthropology* Kant usually spoke of a three-staged educational program, namely: civilization, cultivation, and moralization. According to Kant it is only through development along the lines of such a program that the human being actually becomes human. The truly human state, humanity, is not a given for Kant, but must be brought about through education and upbringing. This education is, on the one hand, the prerequisite for aesthetic experience and judgment. On the other hand, aesthetics or, more precisely, beautiful art, for its part, promotes the development of humanity. This interrelation now needs to be described in detail.

Insofar as the civilizing process is not equated with that of cultivation, but is rather to be seen as the first stage of the program of education, the civilizing process basically means discipline, formation of the ability to work and the capacity for distance. In this sense it is an elemental prerequisite for aesthetic experience. This is not elaborated by Kant in the *Critique of Judgment*, but can be recognized without difficulty. The decisive prerequisite for the aesthetic judgment of objects is the ability to make the distinction between "pleasant" and "beautiful." That assumes, however, that one does not merely abandon oneself to pleasure, but rather that one can preserve distance and judge disinterestedly things as such, on the basis of their qualities. This can be studied well in the two vague examples "table decoration" and "background music," already cited. In the paragraphs in which these appear, Kant attempts to distinguish between pleasant and fine arts. The vagueness of the examples, however, shows that he does not really succeed. For whether background music – we might think of the compositions of Haydn – entertains "merely as an agreeable noise fostering a genial spirit" obviously depends, not on the art, but on whether there is "anyone paying the smallest attention to the composition." Background music as well as table decor might very well be judged on the basis of its beauty. It all depends on distance. Namely that one does not experience direct, but only reflective pleasure. This becomes clear at the close of the paragraph mentioned, in which it says:

The universal communicability of a pleasure involves in its very concept that the pleasure is not one of enjoyment arising out of mere sensation, but must be one of reflection. Hence aesthetic art, as art which is beautiful, is one having for its standard the reflective judgment and not organic sensation.[40]

The distance required here between sensuous reception and pleasure drawn from reflection already assumes an initial step of education, that of the civilizing process.

The cultivation of the faculties of feeling (*Gemütskräfte*) also plays a decisive role for the universality of the judgments of taste. If one asks upon what Kant bases his claim of universality in judgments of taste, then the first correct, albeit insufficient response is surely: on the basic common constitution of human beings. In fact, Kant does refer to such an anthropological invariant:

> Judgment can only be directed to the subjective conditions of its employment in general ... and so can only be directed to that subjective factor which we may presuppose in all men (as requisite for a possible experience generally), it follows that the accordance of a representation with these conditions of judgment must admit of being assumed valid *a priori* for everyone.[41]

But the reference to the same basic constitution of human beings is obviously not sufficient, as the continuation of the quoted section immediately demonstrates. Here it reads:

> That is, we may rightly impute to everyone the pleasure or the subjective purposiveness of the representation for the relation between the cognitive faculties in the act of judging a sensible object in general.[42]

"I impute to everyone": that is to say, judgments of taste, strictly speaking, are not necessarily valid statements, but justifiable imputations. On the foundation of the given constitution of the human being it is by no means guaranteed that everyone will arrive at the same judgment with regard to an object, but only that it can be expected of him or her to develop in such a way so as to be sensible to the beauty of this object. The judgment "x is beautiful" made as an utterance to a companion is an "invitation" to put him or herself in a position to share this judgment with me. Conversely not everyone is in the position to pass judgments of taste. Taste is rather something that must be formed. This process of education can be described more specificity as cultivation. Kant here often speaks of the culture of the mind (*Gemütskräfte*). It occurs principally through examples of fine art. Thus it is said for instance of the artist that he has "practiced and corrected his taste by a variety of examples from nature or art." In other places Kant speaks of a "culture of the mental powers" and of the fact that "we estimate the worth of the fine arts by the culture they supply to the mind."[43]

Kant's conception that taste is a matter of education is most apparent where, in a fashion typical of eighteenth-century Eurocentrism, he sets the aesthetic culture of those who are civilized apart from that of the primitives. The section in question comes from the paragraph on the empirical interest in the beautiful, which I have already quoted, but which I cite here again in full:

> With no one to take into account but himself, a man abandoned on a desert island would not adorn either himself or his hut, nor would he look for flowers, and still less plant them, with the object of providing himself with personal adornments. Only in society does it occur to him to be not merely a man, but a man refined after the manner of his kind (the beginning of civilization) – for that is the estimate formed of one who has the bent and turn for communicating his pleasure to others, and who is not quite satisfied with an object unless his feeling of delight in it can be shared in communion with others. Further, a regard to universal communicability is a thing which everyone expects and requires from everyone else, just as if it were part of an original compact dictated by humanity itself. And thus, no doubt, at first only charms, e.g., colors for painting oneself (roucou among the Caribs and cinnabar among the Iroquois), or flowers, sea-shells, beautifully colored feathers, then, in the course of time, also beautiful forms (as in canoes, wearing-apparel etc.) which convey no gratification, i.e., delight of enjoyment, become of moment in society and attract a considerable interest. Eventually, when civilization has reached its height it makes this work of communication almost the main business of refined inclination, and the entire value of sensations is placed in the degree to which they permit of universal communication. At this stage, then, even where the pleasure which each one has in an object is but insignificant and possesses of itself no conspicuous interest, still the idea of its universal communicability almost indefinitely augments its value.[44]

This section describes the civilizing process as the development of taste. It leads from the pleasant to the beautiful and out of the isolation of private enjoyment into the community of "refined human beings," who are in the position to mutually partake of one another's sensibility through the selection or arrangement of things. Philippi, attending Kant's *Logic lectures*, noted accordingly: "Taste calls for cultivated, sensuous judgment" and "Sociability demands that we be able to judge what will please our friends."[45]

Insofar as the cultivation of taste is already a surmounting of the private and the empirical interest in the beautiful is directed towards the development of community, Kant views this step in the development of humanity as already a transition to morality (*Sittlichkeit*). This is explicitly necessary for the next stage, that of "aesthetic maturity," if it may be expressed this way, namely in order to be able to take an intellectual interest in the beautiful. This interest, as demonstrated in §42, is based on an analogy between a feeling for the beautiful and moral feeling. The moral feeling of respect appears when practical maxims "are of themselves qualified for universal legislation."[46] Corresponding to this in the realm of aesthetics is the pleasure in the fact that nature, of its own accord, appears to reveal itself to our cognitive capacity. To be able to appreciate this analogy, one must already have educated one's moral consciousness. Kant says: "This immediate interest in the beauty of nature is not in fact common. It is peculiar to those whose habits of thought are already trained to the good or else are eminently susceptible of such training." Shortly before he speaks of "the refined and well-grounded habits of thought of all men who have cultivated their moral feeling."[47]

In a later passage Kant will describe "the development of moral ideas and the culture of the moral feeling" as "the true propaedeutic for laying the foundations of taste"[48] but the development of moral consciousness is yet only the precondition for the *intellectual interest* in the beautiful. In the case of the sublime, admittedly, the situation is different. Here the *experience* of the forces of nature as sublime is possible only on the basis of an already developed moral consciousness.

> In fact, without the development of moral ideas, that which, thanks to preparatory culture, we call sublime, merely strikes the untutored man as terrifying. He will see in the evidences which the ravages of nature give of her dominion, and in the vast scale of her might, compared with which his own is diminished to insignificance, only the misery, peril, and distress that would compass the man who was thrown to its mercy.[49]

Only those, whose course of development has already attained this stage of moralization, are receptive to the sublime. The sight of the magnitude and the violence of nature will release in them a feeling that they themselves belong to a wholly other world, to the intelligible world. The path of the aesthetic education leads, as Schiller quite rightly perceived, to the consciousness of freedom.

Conclusion

We began with the difficulties experienced by the contemporary reader of Kant's *Critique of Judgment* and with the dissatisfaction generated by the state of current Kant research. Both result, in my view, from adhering too closely to the method of reading suggested by the logic and didactics of Kant's text, namely: that it deals with the foundation of aesthetic judgments or, expressed in modern terms, of aesthetic discourse. The genuine richness of the work only unfolds, when attention is paid to what it indirectly communicates, through examples, footnotes, and excurses. In truth it is a highly sensitive and sympathetic examination of the phenomenon of beauty and of the experience of beauty.

Beauty is a typical "intermediary phenomenon." It cannot be expressed in objective discourse, that is, through the statement of the particular properties of the object, nor is it simply a projection of the states of the subject. Beauty is something atmospheric, it is the characteristically perceptible presence of objects and at the same time the quasi-objective condition of subjects.

If one reads Kant's *Critique of Judgment* as a theory of beauty in this way, one might well be disappointed in the face of the "innocuousness" of the experience of the beautiful evident in it. To repeat once again: beauty for Kant is neither something erotic nor something terrible, neither something divine nor an overpowering force. It would therefore be quite out of place to read Kant's text as *the* theory of beauty. One does Kant the most justice if one refrains from coming to him with such totalizing intentions, inspired by veneration for the philosopher. As fundamental as is the work of this philosopher, he nevertheless was also limited as a man and thinker of a particular epoch. From a precise observation of Kant's examples in his aesthetics, we came to the conclusion that Kant's aesthetics is the adequate theory of the aesthetic sensibility of the rococo and the cultivated bourgeois style of life in the second half of the eighteenth century. This historicization of Kantian theory in no way diminishes its significance and its truth. It is rather the case that the aesthetic experience of the cultivated bourgeoisie constitutes one possible form, one could say perhaps one dimension, of experiencing the beautiful, which belongs fundamentally to the phenomenon of beauty. We would grant the same to Plato's theory: even if it is necessary to understand that the experience of beauty as something divine derives from a particular historical context, there is, nevertheless, something fundamental in this experience, which remains a permanent potential of the phenomenon of beauty. The same applies to Kant's theory of beauty. From the phenomenon of beauty Kant constructed the sociability of feelings. The relationship between the phenomenon of beauty and communication and education assigns to beauty a social function: beauty serves the cultivation of the human into a social being.

Notes

1 Translated by Andrea Lobb.
2 J. W. von Goethe, "Anschauende Urteilskraft," *Goethe's Werke*, Hamburger Ausgabe, vol. XIII, 30, pp. 30–7.
3 This and the following quotations of Kant are reproduced here from the standard English translation, James Creed Meredith's *Kant's Critique of Aesthetic Judgment*, Oxford, Clarendon Press, 1911.
4 H. Vaihinger could therefore present a large portion of Kant's work in his book *Die Philosophie des Als Ob*, Leipzig, Meiner, 1927.
5 Kant, *Critique of Aesthetic Judgment*, pp. 104–5.
6 Ibid., pp. 57–8.
7 Ibid., p. 140ff.
8 Ibid., p. 56.
9 In my essay "Atmosphere as the Fundamental Concept of a New Aesthetics," *Thesis Eleven*, 36, 1993, 113–26.
10 Kant, *Critique of Aesthetic Judgment*, p. 130ff.

11 Ibid., p. 41.
12 Ibid., p. 46.
13 Ibid., p. 52.
14 Ibid., pp. 86, 89.
15 Particularly J.- F. Lyotard, *Leçons sur l'analytique du sublime*, Paris, Galilée, 1991. See also C. Pries (ed.), *Das Erhabene. Zwischen Grenzerfahrung und Grössenwahn*, Weinheim, VCH, 1989.
16 K. Vorländer, *Immanuel Kant, Der Mann und das Werk*, 3rd edn, Hamburg, Meiner, 1992, Book 3, Chapter 5, "Kant und die Kunst."
17 Kant, *Critique of Aesthetic Judgment*, pp. 70, 88.
18 Ibid., pp. 46, 162, 155, 158.
19 Ibid., p. 72.
20 Ibid., pp. 166, 187.
21 I quote here the remarks on aesthetics, which are found in Philippi's transcript of Kant's *Lectures on logic*. Quoted in J. Kuhlenkamp (ed.), *Materialien zur "Kritik der Urteilskraft,"* Frankfurt/M., Suhrkamp, 1974, p. 101ff.
22 Vorländer, *Immanuel Kant*, Book 2, Chapter 2.
23 L. F. Borowski, *Darstellung des Lebens und Charakters Immanuel Kants*, Darmstadt, Wissenschaftliche Buchgesellschaft, 1968, p. 56. It is indicative that Garve in his *On fashions* (a treatise obviously written in direct response to the *Critique of Judgment*) wants to limit Kant's analysis of the beautiful to those objects subject to fashion. C. Garve, *Versuche über verschiedene Gegenstände aus Moral, Litteratur und dem gesellschaftlichen Leben*, Part 1, Breslau, W. G. Korn, 1792. He distinguishes there between objective and subjective beauty, while viewing the first – in opposition to Kant – as grounded in natural given form.
24 F. Schiller, *Über die ästhetische Erziehung des Menschen in einer Reihe von Briefen*, with a commentary by Stefan Matuschek, Frankfurt/M., Suhrkamp, 2009 [1795].
25 J. Burke (ed.), W. Hogarth, *The Analysis of Beauty*, Oxford, Clarendon Press, 1955, pp. 55–6.
26 Ibid., p. 37.
27 Ibid., p. 42.
28 In the epoch of sensibility, the *"je ne sais quoi"* became the set phrase to express that one was moved. It appears, however, that it was already an established *topos* of courtly aesthetics. Montesquieu, in his essay *Du je ne sais quoi*, printed under gout in the *Encyclopédie*, names the *"je ne sais quoi"* as an object of taste beside the beautiful, the good, the agreeable, the naive, the delicate, the gentle, the gracious, the noble, the great, the sublime, and the majestic. Characteristic of the *"je ne sais quoi"* according to Montesquieu, is that it is a surprising charm, one that was not to be expected according to normal standards of beauty, splendor, etc. He therefore connects it with natural grace.
29 Kant, *Critique of Aesthetic Judgment*, p. 150.
30 Ibid., p. 154.
31 Ibid., p. 180.
32 Ibid., p. 194.
33 *De signatura rerum*. See also my essay *Sprechende Natur: Die Signaturenlehre bei Paracelsus und Jacob Böhme. Für eine ökologische Naturästhetik*, Frankfurt/M., Suhrkamp, 1993, 2nd edn.
34 Kant, *Critique of Aesthetic Judgment*, p. 42, my emphasis.
35 Ibid., p. 155.
36 *Lectures on Logic*, in Kuhlenkamp (ed.), *Materialien*, n. 7, p. 107ff. It is of course clearly evident from the formulations that this is not Kant himself speaking here.
37 E. Burke, *A Philosophical Enquiry into the Origin of our Ideas of the Sublime and Beautiful*, Notre Dame, University of Notre Dame, 1958.
38 Kant, *Critique of Aesthetic Judgment*, p. 156.

39 Ibid.
40 Ibid., p. 166.
41 Ibid., pp. 146–7.
42 Here the translation of S. H. Bernhard (1892) is preferable. *Critique of Judgment* (New York, 1951), p. 132.
43 Kant, *Critique of Aesthetic Judgment*, pp. 174, 226, 195.
44 Ibid., pp. 155–6.
45 Ibid., pp. 96–7.
46 Ibid., p. 156.
47 Ibid., pp. 160, 158.
48 Ibid., p. 227.
49 Ibid., p. 115.

5   On beauty

The riddle of beauty

That we are touched by beauty is beyond doubt. Everyone has such experiences, and knows that they cover a wide spectrum. They extend from deep shock, through amazed perception of something wholly other, to a momentary lifting of our spirits. They extend from a feeling of painful longing, through fascination with the wholly other, to a sense of being enveloped and sheltered in delightful well-being. Yet if we are called upon to say what beauty is, we find ourselves at a loss. Indeed, we are afraid that by naming beauty we may turn it into something other than what we have experienced.

> We cannot possibly be inclined to comprehend in the form of a definition something which must be understood in a fundamentally different way, something which we have ourselves experienced and loved quite differently; for such a definition can easily make it something alien and different.[1]

With this assertion the philosopher and religious thinker Søren Kierkegaard introduces the concepts of "existence." These are concepts with which matters become *serious*, because the content of the concept is intimately bound up with the person who thinks it, or, conversely, because the subject is involved in the content of his or her thought. Perhaps we have already discovered here a first characteristic of beauty. Beauty cannot be completely objectified; beauty is not a property which a person, a thing, or a scene simply *has*, because the involvement of the participating subject is always intrinsic to it. We cannot pin down the beauty of a scene just because our delight is an intrinsic part of that scene. One might leave the matter there and say: *One cannot speak of beauty, one has to experience it.*

But one does not leave the matter there. On the contrary, since the ancient Greeks there have constantly been attempts to determine what beauty is,[2] and in philosophy a whole discipline has emerged which is primarily concerned with this question, namely aesthetics. Up to the mid-nineteenth century aesthetics was a theory of beauty. It might also have been called a theory of the fine arts, since up to about that time the essential demand placed on the work of art was that what it represented had to appear beautiful. A hint of this view still persists in our everyday notions, as when we say of a thing or a situation that it is *aesthetically pleasing*, by which we mean that it is beautiful; or when we speak of the *aestheticizing of the real*, referring to attempts to beautify reality. It was only after Hegel that something like an *aesthetic of ugliness*[3] could emerge, and, as we know, since then art has moved very far from having to be beautiful as such. And yet it is precisely this aspect, this way of gaining access to the beautiful, by way of art, design, or architecture, which is a further reason for attempting to determine what beauty is. For although the experience of beauty may be, for the recipient, so subjective, indeed so intimate that one cannot hope to elucidate it verbally in its ultimate nature, the matter looks quite different from another direction, when viewed from the standpoint of production aesthetics. It is, of course, the aim of the artist, the designer, the architect, to create by means of the object the conditions in which people are able to experience beauty – by establishing vistas, by shaping objects, by arranging scenes. This implies two things. First, that the experience of beauty cannot be as subjective as it first appears to the person affected by it. If the productive effort to create beauty is to have any meaning at all, then it must be supposed that our experiences of beauty are, at least to a certain extent, shared. The paradigm here is once again – as in the aesthetics of atmospheres generally[4] – the art of the stage set. That art would be pointless if it could not be assumed that a given audience would experience in the same way an arrangement with which an atmosphere is created on the stage. It is the same with beauty: the artist, the designer, the architect will want to know what he or she has to do to ensure that a public will experience his or her objects or arrangements as beautiful. And to say what the artist has to do would be the task of aesthetics.

And yet, has it been said? Have we got a definition of beauty? Or must it constantly be defined anew, because people's taste changes, or perhaps – and this goes deeper – because people's manner of experiencing changes?

The classical answers

Just as the art and architecture of the ancient Greeks have been the dominant influences over the longest period of European cultural history, so too have their views on beauty. At the beginning of any aesthetic theory stands Plato, with this thesis that beauty was that which shines forth most strongly (το εκφανεστατον).[5] This definition has something in its favor even today. It seems to locate the beautiful in the sphere of the visible. It brings it together with the experience of light, and although it defines beauty very clearly as a characteristic of objects, it nevertheless, by using the term "shine" (φαινεσθαι), establishes a relationship to the subject: if something appears (φαινεθαι), it must do so for someone. Yet this first impression is deceptive; in it we read Plato's definition too quickly, in terms of our own needs. First of all, it must be remembered that for Plato true being, which also means true beauty, lies in the forms (or ideas), that is, not in the things of the sensuous world but in the original models existing in eternity, which, as such, can only be *thought*. The things in the world are beautiful only in so far as the *Forms* are brought clearly to expression through them. This makes clear at the same time what the reference to "that which shines forth most strongly" actually means. Just as the eternal things, the *forms*, are beautiful in being simply what they are, the transient things of our world are beautiful in making clearly visible the form to which they correspond. A bed is beautiful by being a good bed, and as such recognizable. Something is beautiful if it is what it is really *well*, and as such is knowable. In this way, beauty becomes identified with a kind of radiant precision.

This view has a certain fascination, and one would like to follow it if one could. But it is clearly based on presuppositions which we no longer share. Plato's conception of the beautiful places the beauty of something – of things, but also of people – in relationship to what they truly are. Beauty, therefore, is in no way something external, a veneer. It is not mere appearance, but is bound up with being-good. In the background of this idea is the conception, shared by all Greeks, of the unity of the beautiful and the good, of the *kalonkagaton*. Thus the good citizens are those held in esteem, and as esteemed citizens they are also beautiful. The beautiful are the aristocrats, who are distinguished by their proficiency, which sets them apart from the mass. Just as being beautiful is not something external to things or people, likewise being-good is not something which has its effect in seclusion. Inconspicuous goodness was a notion alien to the Greeks. What is good – such was their conviction – also reveals itself as such. It is radiant, it stands out. Socrates, in the Platonic dialogue *Hippias Major*, brings both ideas together very beautifully when the question is raised whether a household mixing implement, a "quirl," made of gold is more beautiful than one made of olive wood. For Socrates the matter is clear. The quirl made of gold may be lustrous, but it is difficult to handle and does not harmonize with the taste of the gruel it is supposed to stir. Therefore, the olive wood quirl is more beautiful, and that is seen both in using it and in the taste of the gruel produced with it. Goodness stands out, and the better a quirl is, the more precise and radiant it is, and therefore the more beautiful.[6]

The linking of being-beautiful and being-good in Plato has far-reaching consequences. It is already clear in his own work that, according to his definition of beauty, mathematical objects, and, more precisely, geometrical ones, are the most beautiful. For nothing is as good, as precisely what it is, as a mathematical object: a sphere, a tetrahedron, a cube, a square, an equilateral triangle. Beauty thus becomes regularity, harmony, proportionality. The consequence – disastrous from a present-day standpoint – is that attempts have been made on this basis to decipher the secret of beauty in other spheres: simple numerical relationships, the Golden Section, arithmetic, and geometric means, the Fibonacci series, have been used to explain what could cause us to experience something as beautiful.[7] And there is indeed something satisfying in noting that the seed head of a sunflower or the convolutions of a snail's shell obey certain mathematical laws, and even in fractal formations such as the *Appelman* (*Mandelbrot-Set in chaos theory*) – which, though confusing, impress us as beautiful – structures of self-similarity can be found. The idea of "beauty through proportionality" has also proved itself in practice. The impression of measure and balance, the graceful beauty of classical buildings from Roman antiquity up to neoclassicism, is based on confidence in the theory of proportion of Vitruvius. This theory, however, apart from recommending that the dimensions and parts of a building be determined by a basic measure and the whole be thus made proportional, also has the peculiarity that the basic measure is the human being itself. It therefore contains the subjective moment which seems to us to be necessary for the experience of beauty: in a Vitruvian building human beings can set themselves in relation to the whole because their own bodies are the measure of that whole.

The aesthetic theory of Immanuel Kant marked a decisive turning point in thinking about beauty. And yet even his thought was dominated by the basic idea that true beauty is located in forms, which was to play a leading role right up to the twentieth century. In Kant the debate on whether line or color was more important, a debate which, of course, took place in relation to painting, was in reality decided in advance, since he only knew the paintings which were regarded as models in his time in the form of steel engravings. He intensified this antithesis by connecting it to the difference between the pleasing and the beautiful. Colors are, at most, pleasing, whereas beauty depends on form, and therefore, in paintings, on line. Accordingly, he said of music that it is "enjoyment rather than cultivation,"[8] since the pleasure afforded by music is based on the associations it arouses. "But in the charm and mental movement produced by music, mathematics has certainly not the slightest share."[9] With Plato music has been a paradigm of beauty precisely because of its mathematical character. Even before him, Pythagoras had represented the octave by the ratio 2:1, the fifth by the ratio 3:2, and the fourth by the ratio 4:3. But between Plato and Kant stands the astronomer Kepler, who had stated that everyone needed to have a mathematician deep in their soul in order to take pleasure in music.[10] Kant no longer wanted to believe in such a mathematician, who, in listening to music, perceived its harmonic proportions. For him, therefore, beauty was no longer a matter of cognition, of perfected form, but rather a matter of the arousal of human emotions in the search *for* form.

This difference does in fact constitute a decisive turning point in thinking about beauty. It meant that emotional, affective participation in beauty was taken seriously. The beautiful, Kant rightly observes, sets the forces of feeling in motion. It is uplifting. For Kant, objects such as the Platonic bodies, cubes, tetrahedrons, octahedrons, spheres, etc., were not beautiful. They were perfect, they corresponded to their concept and for that reason could be thought, but the thought of them was not connected to a feeling. By contrast, forms that have a certain indefiniteness, of the kind which stimulates the human imagination, which cause us to ask what they might be, are to be called beautiful. The joy one feels in contemplating them in their presence, is, according to Kant, the joy in one's own emotional activity, in the agitation and play of the emotions in their search for form.

For modern people, this view may well have something to recommend it. It does justice to the subjective element in the experience of beauty. *It is not the – sober – noting of the perfection of a form which constitutes the experience of beauty, but rather the searching and tracing-out that a form induces.* Kant was a man of the rococo; what was aesthetically relevant for him in the everyday world was its playful forms, the resonances, and suggestions in a decor. The basic form, for this aesthetic, was not the sphere or the tetrahedron, but the seashell profile or, as in the graphic work of Hogarth, the unbalanced S-shape. The latter's book *The Analysis of Beauty*[11] contains a table of model drawings, including a line of beauty which corresponds approximately to the leg of a baroque chair or the corseted form of the upper part of a female body.

It might be said that Kant brought beauty down from heaven to earth. If for Plato beautiful things are the eternal Forms, for Kant they are the objects of everyday life. There has been much dispute over whether Kant's aesthetics is an aesthetics of art or of nature. But if one looks more closely one finds that the majority of his examples are taken from the field of design. He is interested in fashion, in tapestry; he speaks of front gardens and of table music.[12] Aesthetic judgment is the ability to choose the right things with which to embellish life. Beauty has a community-forming function. In the pleasure taken in certain things, people with the same taste feel united. Through the furnishing of one's surroundings one creates for oneself and others a stimulating, enlivening atmosphere. That Kant still saw the reason for this primarily in the forms of objects may be a limitation; nevertheless, he liberated thinking about beauty from the domination of mathematics. Even when he is concerned with form, it is not consummate form, it is not perfection and precision, which today have succumbed beyond recall to the triviality of industrial production; rather, it is the degree of indefiniteness in form, the suggestive, the vague, the playful, which occasion the experience of beauty. In this, Kant points far beyond his own work.

New experiences

It is a bold assertion to state that people in different historical periods perceive in a different way. And yet modes of perception mediated by technology – in the visual sphere, since the invention of the telescope and the microscope – are likely to represent at least an enlargement of the field of perception and perhaps even fundamental changes.[13] That will also have consequences for our conceptions of the beautiful. All the same, how one thinks about beauty is very fundamentally influenced by paradigmatic objects. No doubt we are far from asserting today that the human body is the paradigm *par excellence* of beauty, as Friedrich Schiller claimed and as was probably characteristic of German classicism as a whole. Still more categorical, it seems, is the rejection of simple forms as paradigms of the beautiful. It is not only that simple geometrical forms bore us – the glass pyramid in the courtyard of the Louvre was, after all, a mistake – but the simple yet sophisticated forms of a Brancusi or a Henry Moore, acclaimed half a century ago, now leave us cold. It seems that what was prefigured in Kant, the charm of the indefinite in form, has become autonomous as indeterminateness as such. One need only think of the fascination exerted by the seascapes of a photographer such as Hiroshi Sugimoto.

Consider water: within the Kantian aesthetic – as in the classical aesthetic generally up to and including that of Adorno – it would have been unthinkable to derive aesthetic charm from water as such. Water is simply the formless, and for that very reason was seen by Aristotle – as that capable of being formed – as the antithesis of all form. A stream, a waterfall, a pond – such a thing is nowhere to be found in the Kantian aesthetic. It is true that he does mention the sea, but not under the heading of *beauty* but of *sublimity*. The sea in its immeasurable vastness and its menacing violence is experienced, from a suitable distance, as sublime. Of course, water or, more precisely, stretches of water, have often been present in landscape painting since its beginnings, but only as an ingredient, a component for generating a landscape-like atmosphere. In this capacity water was then explicitly appreciated by C. C. Hirschfeld in his theory of landscape gardening.[14] He inquired into which type of body of water was suited to which natural scene, or, more precisely, what a certain body of water contributed to the emotional character of a scene. From there it seems only a step to the paintings of Turner, who sought to paint atmospheres as products of the interplay of water, light, and weather – in which interplay the world of objects receded entirely. Yet this was a large step, and it appears to have been taken by the wider public only 100 years later.

All this was anticipated theoretically, of course, in Goethe's theory of colors, which had ascribed to colors *a sensuous and moral effect*. Today we would speak of an atmospheric effect. Colors communicate to space a certain mood. And because this mood is apprehended by a person present in terms of feeling, the interplay of colors in their vividness can be experienced as beauty. This effect, as far as it can be mediated by panel paintings, is experienced most clearly in the works of Mark Rothko.[15] However, one should not contemplate them with the intention of ascertaining what can be seen in them; rather one should allow their color effects to unfold spatially. If one starts out from the old antithesis of color versus line, the paintings of Mark Rothko might be regarded as the new paradigm of beauty.

But, in this context, that does not take us very far. When a new idea of beauty is at issue we are concerned, of course, with the experience of light, water, space, and color as such, and with the pleasure of indefiniteness, of which Kant said that it stimulates the imagination. But it must be said first

62 *Theory: aesthetics and aesthetical economy*

that it was new basic examples of beauty, new paradigms of perception, which prepared the way for this concept of beauty. And these have been mediated largely by the development of technology.

This interconnection between technological development and the change in conceptions of beauty has two sides. On the one hand, the basic technical conditions of experience have made new perceptual pleasures available to people in the modern period. On the other, the technical mastery of light and sound, together with the technical shaping of material or, still more, of materiality,[16] have made possible the generation of practically unlimited aesthetic effects. The change in perceptual experience may have begun with the microscope – less for the scientist, who saw something specific through it, than for the lay person, who gazed through the microscope into unknown worlds. In this context Ernst Haeckel's radiolaria undoubtedly still belongs to the old paradigms of beauty.[17] By contrast, flickering forms at an indeterminate depth are something really new, especially when they appear in the fascinating illumination of polarized light. Closest to these are the experiences one has when diving, or through the mediation of underwater photography. Here again we find the indeterminateness, above all the floating sensation and the lighting, but most especially, in this case, the perception of the medium as well. It gives a sense of being present in a way which would never be possible in an object-determined world. Then there are the experiences of flying, which by now can be assumed to be shared by almost everyone. Here, one finds oneself in a space of flickering forms which change their appearance, depending on illumination and position. The gaze stretches over infinite vistas with contours always unique, ephemeral. One encounters here not only a diffuse infinity but an almost surreal clarity of formations, with ungraspable contours. It is no wonder that clouds, which were always to be seen creating aerial perspectives in landscape paintings, should also, like water and light, have made themselves autonomous in art. It was no doubt Alfred Stieglitz who began this tendency about 1900, with pure cloud photography. This theme is ubiquitous today, and in Richter has been extended to panel painting.[18] Admittedly, in these object-fixing arts, much of the original aesthetic fascination of clouds is lost. In a photograph they are easily reduced to a thing; with their mutability they also lose the charm they had for the imagination.[19]

What is crucial here is that such experiences can be assumed to be part of the general stock of experience. They belong to the *normal* spectrum of perception. For many people, further experiences, those of explicitly artificial worlds, can be added to this spectrum. One thinks of drug experiences, in particular of LSD, an experience of weightlessness which is generated by this or other means, or experiences of computer games or other virtual worlds. What matters here is whether a person is merely confronted with this world as an image, or experiences it as a space in which he or she is physically present. Only in the latter case can one speak of virtual worlds in the strict sense.[20] But whether or not technical devices mediate these experiences in a particular case, they have an inherent tendency to abolish the division between dream and reality, a division by which the modern age was once defined.[21] And precisely this appears to be the objective of postmodern pop aesthetics. It simulates worlds in which one is present in the form of an avatar: cave experiences in which one is present in simulated surroundings by means of data gloves and electronic spectacles. Or, inversely, it makes possible bodily presence in a simulated scene, like the one which can be experienced by a visitor to a casino in Las Vegas, who is catapulted by a lift into a scene from *Star Trek*.[22] This is the point where we have to speak not only of new experiences, but of new needs.

New needs

It cannot be said that, in the light of these new experiences, the classical paradigms of beauty have been simply devalued. But we see them in a new way, we have new expectations of them; they practically have to prove themselves once again in experience. We would not be satisfied to know that they are perfect according to this or that criterion; we will ask what, or, better, whether we feel something in their presence. Experiences such as the *Venus de Milo* or Michelangelo's *David*, Cologne cathedral, the Sforza Castle in Milan, and the Alhambra in Granada would certainly pass this test. It is just that the spectrum has widened considerably, and we will describe as beautiful quite different things and scenes than were possible according to the classical aesthetic theories – a spider's web glistening with raindrops, a sun-steeped avenue of trees, but also the design of the illumination in an underground station or the sales display in a high-class departmental store. Today, beauty can no longer be banished to the museum, it is no longer defined by the difference between serious and popular art. In principle, we look for beauty everywhere. We can find beauty not only in art but also in advertising, in design, in urban scenes, in nature, and in the artificial worlds of our media. The only thing that counts is the quality of the impression emanating from a person, a scene, an object, a piece of architecture. What is decisive for us today, when we use the word beauty, is whether a person or a thing, a scene or a place makes us feel that we are there, whether these things, people, or scenes contribute to intensifying our existence.

This enables us to define once again, and in conclusion, the difference from classical notions of beauty. Plato, for good reasons, brought together beauty and Eros. Eros, love, he thought, was the desire to possess the beautiful, and then, still more trenchantly, the desire *always* to possess the beautiful.[23] Although we can still empathize with this idea, its weaknesses are undeniable: for if love is the desire to possess the beautiful, it will only remain alive for as long as one does not possess beauty, or for as long as its possession is at risk. What is more important, however, is the assumption contained in this relating of desire to the beautiful – that beauty as such is something lasting. And for Plato beauty is indeed ultimately an eternal *Form*, and is present in the sensuous world only in a highly fractured way. A corresponding assumption underlies the traditional striving of artists to create works, that is, something permanent. We, by contrast, have become more modest or, better, more sensuous. We are able to experience beauty in the ephemeral, the transient, in the light glinting on a pewter vessel,[24] or in the play of shadow on a white wall. Because we ourselves are transient beings, we encounter beauty in the lighting-up of appearances which assure us of our existence. *Beauty is that which mediates to us the joy of being here.*[25]

Notes

1　Søren Kierkegaard, *The Concept of Dread*, translated from *Der Begriff Angst*, Reinbek, 1960, p. 133f.
2　Michael Hauskeller, *Was das Schöne sei. Klassische Texte von Platon bis Adorno*, Munich, Darmstadt, 1994.
3　Karl Rosenkranz, *Ästhetik des Hässlichen. Mit einem Vorwort zum Neudruck von Wolfhart Henckmann* [reprographic reprint of the Königsberg edition of 1853], Munich, Darmstadt, 1989.
4　Regarding my attempts to treat beauty itself as atmosphere, see the chapter "Schönheit und andere Atmosphären," in Gernot Böhme, *Anthropologie in pragmatischer Hinsicht. Darmstädter Vorlesungen*, Frankfurt/M., Suhrkamp, 1994, and the chapter "Die Gegenwart des Schönen und die Kultivierung des Lebens," in Gernot Böhme, *Kants Kritik der Urteilskraft in neuer Sicht*, Frankfurt/M., Suhrkamp, 1999; translated as *Kant's Aesthetic: A New Perspective, Thesis Eleven*, 43, 1995, 100–19.
5　Plato, *Phaedrus*, 250D.
6　Gernot Böhme, "Der Glanz des Materials: Zur Kritik der ästhetischen Ökonomie," in *Atmosphäre: Essays zur Neuen Ästhetik*, 6th edn, Frankfurt/M., Suhrkamp, 2009.
7　Bernd-Olaf Küppers, "Die ästhetische Dimension natürlicher Komplexität," in Wolfgang Welsch, *Die Aktualität des Ästhetischen*, Munich, Fink, 1993, pp. 247–77.
8　Immanuel Kant, *Kritik der Urteilskraft*, Hamburg, Felix Meiner, 1959, p. 185 (original edn. of 1790, p. 218), English translation by J. H. Bernard: *Critique of Judgment*, London, Collier Macmillan Publishers, 1951, p. 172.
9　Ibid., p. 186 (orig. edn. p. 220); English translation, p. 173.
10　Gernot Böhme, "Von der Sphärenharmonie zum Soundscape," in Gerhard Kilger (ed.), *Macht Musik. Musik als Glück und Nutzen für das Leben*. Catalogue of the exhibition of DASA, Bundesanstalt für Arbeitsschutz und Arbeitsmedizin, Dortmund/Cologne, Wienand Verlag, 2006, pp. 30–7.
11　William Hogarth, *The Analysis of Beauty. Written with a View of Fixing the Fluctuating Ideas of Taste*, London, John Reeves, 1753.
12　See the list of examples relating to aesthetics in the *Critique of Judgment* in my book *Kants "Kritik der Urteilskraft" in neuer Sicht*, Frankfurt/M., Suhrkamp, 1999. This list was first published in Danish: Gernot Böhme, "Index over de aestetiske exempler in Kants *Kritik der Urteilskraft*," *Kritik*, 105, 1993, 79–80.
13　Gernot Böhme, "Technisierung der Wahrnehmung: Zur Technik- und Kulturgeschichte der Wahrnehmung," in J. Halfmann (ed.), *Technische Zivilisation: Zur Aktualität der Technikreflexion im der gesellschaftlichen Selbstbeschreibung*, Opladen, Leske und Buderich, 1998, pp. 31–47.
14　Gernot Böhme, "Ästhetik der Gewässer," in Annegret Laabs and Eckhart W. Peters (eds.), *Die Elbe [in] between*. Symposium, Magdeburg, Kunstmuseum Unser Lieben Frauen, 2007, pp. 62–9.
15　In contrast, the work of his fellow abstract expressionist, Barnett Newman, stand for the opposite classical concept, the sublime.

16 On this distinction, see Böhme, "Der Glanz des Materials."
17 Ernst Haeckel, *Kunstformen der Natur*, Leipzig und Wien, Verlag des Bibliographischen Institus, 1914.
18 S. Kunz, J. Stückelberger, and B. Wismer (eds.), *Wolkenbilder: Die Erfindung des Himmels*, Munich, Hirmer, 2005.
19 At the beginning of Act IV of Goethe's *Faust*, for example, Faust glimpses, in the cloud with which he has returned from Arcadia, Juno, Leda, Helen, and even, in the rising mist, Gretchen.
20 Gernot Böhme, "Der Raum leiblicher Anwesenheit und der Raum als Medium von Darstellung," in Sybille Krämer (ed.), *Performativität und Medialität*, Munich, Fink, 2004, pp. 129–40. English version in Ulrik Ekman (ed.), *Throughout: Art and Culture Emerging with Ubiquitous Computing*. Cambridge, MA, MIT Press, 2010.
21 Stefan Niessen, *Traum und Realität: ihre neuzeitliche Trennung*, Darmstadt, Technische Hochschule, Diss., 1989.
22 Natascha Adamowski, *See You on the Holodeck! Morphing into New Dimensions*: www.ifs.tu-darmstadt.de/fileadmin/gradkoll/Publikationen/space-folder/pdf/Adamowsky.pdf
23 See Diotima's speech in Plato's dialogue *The Symposium*.
24 That was the illumination of the philosopher Jacob Böhme. See my essay "Ästhetik des Ephemeren," in Gernot Böhme, *Für eine ökologische Naturästhetik*, Frankfurt/M., Suhrkamp, 1989, 3rd edn, 1999.
25 Gernot Böhme, "Das Glück, da zu sein," in R. J Kozljanič (ed.), *II. Jahrbuch für Lebensphilosophie*, Munich, Albunea Verlag, 2006, pp. 209–18. Also in Renate Breuninger (ed.), *Glück*, Ulm, Humboldt Studienzentrum, 2006, pp. 57–69.

6 On synesthesia

Of an incidental nature?

People are familiar with *synesthesia*, but they don't take them seriously; they may talk about them, but they do so in a derogatory manner. The idea of a note being soft is somehow incidental, an added fact, or maybe a relict or something indicating a specified origin. Do we really experience things in this manner, or does the derogatory manner of considering *synesthesia* have repercussions on experience? To talk of a soft note is an example of metaphorical speech. In this case an expression from one sensual domain, i.e. that of touch, is transferred to another, i.e. that of hearing. Actually fabrics are soft, and a note is either high or low, or perhaps loud or quiet, long or short. But wait a minute: in what respect is a note high or low? Aren't high and low metaphorical expressions derived from the domain of space, so that, in fact, plains are low and mountains are high? In our conception of music, we are accustomed – primarily owing to the predominance of the piano – to conceiving of all imaginable notes in a row. This establishes an order, both from greater to lesser and an equality. But why do we call one direction of progressing in this order the direction toward the lower notes, and the other the direction to the higher? Couldn't we reverse these concepts? "*Im tiefen Keller sitz ich hier*" we hear the bass voice sing – and it fits. Couldn't it also sing "*Im hohen Ausguck sitz ich hier*"? And which register is suitable for "*Hoch auf dem gelben Wagen*"? A baritone would probably be best. High and low were supposed to be the characteristic terms representing the different notes. We recall that they are not "characteristic." Are there, in fact, any predicates for the domain of music? And conversely: do the terms high and low really characterize the essence of the domain of music? Anyone with a humanist education will recall the mild shock at having to understand that the old Greek expressions for the order relationship in music, which we refer to as high and low, were called *oxys* and *barys*. What we call high was, in classical Greece, sharp (pointed) and what we call low was heavy (weighty) in their language. Did the Greeks experience music differently to us? And how did it come about that the two poles of divergent musical progression are given expressions which derive from completely different realms of experience? How can heavy and sharp be, strictly speaking, the poles of an antinomy? Ordering the notes in accordance with the antinomies of heavy and sharp appears paradoxical at first sight. But once one accepts the idea, doesn't it become plausible? Just imagine a muffled and massy note, weighty and voluminous, its energy concentrated toward the bottom, then imagine how it becomes concentrated, forcing its way upwards – upwards, it should be noted – and initially developing a powerful and then increasingly slender shaft, and in rising further toward the top gradually becoming sharper and more pointed, perhaps even diverging into a number of points. Couldn't this image also represent the order of musical notation? Maybe it would be more adequate than a pure series running from high to low, in which the notes are not distinguished from one another in accordance with weight and character. Having performed these exercises, one will recall that our listening habits have changed as a result of modern music and, above all, modern recording techniques. When music was first reproduced technically by the gramophone record and radio, it was believed that it would be reduced to its pure structure, i.e. to melody and harmony. The other "channels," the mellowness, the individual character of the instruments, the "spiritual" aspect of a voice and the atmosphere of a concert, would be lost through technical reproduction. Meanwhile we have learned that the opposite is the case. The emancipation of the musical avant-garde from any kind of system in music has considerably extended the musically significant qualities of the notes, or let us say in more general terms, their sounds. It is by no means a question of how high a note is, nor of the relevant duration and harmonic significance, but precisely its individual character and perhaps the buzz, the voluminousness, the spatial quality, the inner dynamics, and the indeterminability of its valence which are of interest to the artist. For the listener the fantastic perfection of acoustic techniques has made it possible to listen to music with a quality, i.e. a multidimensionality, that would have never been possible in a concert hall. The individuality of an instrument, the spatial development of a note, its mellowness, atmosphere, and sound are not musically relevant in spite of modern techniques but because of them. Never before has it been possible to hear the human voice in such purity, and so close, as it is now.

A preliminary excursus into one of the sensual spheres, i.e. that of acoustics, shows that the real relations may be precisely the opposite of that which our school knowledge or our common prejudices tell us. It does not appear to be at all certain whether this difference between the qualities, which is supposed to be classified under one sensual sphere, and the *synesthesia* somehow arising elsewhere, is really valid. For the person working in the field of aesthetics (aesthetic workers) this difference may never have existed.[1] Aesthetic workers are here used to mean people who produce for human sensuality. This includes painters and decorators as well as interior designers, musicians, and the salespeople creating the atmosphere in a supermarket, cosmeticians as well as stage designers. For those working in these fields it has always been a question of the synesthetic effect on people's moods. When an interior decorator lays a sea-green carpet in a room, he is not concerned with producing walls with this color but with creating a spatial atmosphere. When a salesperson in a super-market plays a certain type of music, he is not concerned with playing a specific work but with creating a mood conducive to promoting sales. Perhaps this is most striking in the ephemeral art of the stage designer. It is not works which are produced here but "scenes," i.e. atmospherically charged spaces in which a drama can evolve. One could learn more about *synesthesia* from these aesthetic workers than from sensory physiology, psychology, or aesthetic theory. The theory generally takes as its point of departure the prejudice that there are five senses with specific sensual energies and specific sensual qualities, and only searches for so-called intermodal qualities or emotional effects of the sensual qualities when proceeding from this precondition.

The reasons for this theoretical situation seem to me to lie in the fact that the phenomena of human sensuality are not generally analyzed from the point of view of perception, but from that of their cause, the so-called stimuli. It may well be time to invert this relationship. Before any such attempt is made that great exception should be considered, that author from whom one can still learn a great deal about research into human sensuality: Goethe.

Goethe

The sixth section of Goethe's *Theory of Colours* deals with the "*sinnlich-sittliche* effect of color." From this heading one could think that Goethe also first believed that color existed for itself and then in its effect. On closer examination this turns out to be untrue. Even in the central part of his *Theory of Colours* he does not make use of the concept of causality, in contrast to Newton's theory of colors. For Newton light and its physical properties come first, and then color sensations are ascribed to light its specifications and mixtures as effects. He tries to grasp these relations by a first attempt at a kind of psycho-physics. For Goethe, by way of contrast, color is, from the very start, a sensual phenomenon as it presents itself to the eye – and then he merely asks after the conditions for the appearance of colored phenomenon.[2] And in the case of the so-called *sinnlich-sittliche* effect, it is not a question of color being taken as a stimulus and then studying its reaction on mood, but its "effect" is its reality per se. The reality of colors is their presence in the sense of sight of the eye. Speaking of them Goethe says that it "immediately attaches itself to the Sittliche" (*Theory of Colours*, Part 758). The expression "*sittliche*" may seem strange to the contemporary reader, but for Goethe it did not signify morality so much as attitudes, a vital awareness and an ethos. With the expression "*sittliche Wirkung*" (*Wirkung* = effect) Goethe points out that the sensual presence of colors has an impact on our vital awareness. The manner in which this happens is not simply a fact of natural history. Rather Goethe is fully aware of the fact that socialization, refinement, and even convention play a role here. This can be seen in particular in "preferences" attached to colors of mourning in different cultures. Below, Goethe is quoted in extenso on the *sinnlich-sittliche* effect of blue.

1 Just as yellow always contains light, one can say that blue always contains something dark.
2 This color has a peculiar and almost inexpressible effect on the eye. As a color it is energy; on its own it is to be found on the negative side and is also, in its greatest purity, a stimulating nothing. It appears as something contradictory, of stimulus and tranquility.
3 Just as we see the heavens high above and the distant mountains as blue, a blue surface also seems to retreat before us.
4 Just as we like to follow a pleasant object which flees from us, we also like to look at blue, not because it imposes itself on us, but because it draws us towards it.
5 Blue gives us a feeling of coldness, just as it also reminds us of shadows. We are aware of its derivation from black.
6 Rooms which are wallpapered a pure blue appear to be somehow extensive but in fact empty and cold.
7 Blue glass shows objects in a melancholic light.
8 It is not unpleasant when blue participates in plus to a certain extent. Sea green is more of a pleasant colour.

I shall merely pick out the insights relevant for our subject: quite remarkable here is Goethe's statement that the color blue is "energy." Does Goethe want to distinguish blue from all other colors, or are there colors which are not energy? One thing is sure: that Goethe experiences colors as sensual beings, which somehow move and captivate human beings. In the case of blue this energy is by no means only accessible to human beings, but rather something in flight. In section 780 Goethe attributes to blue a certain form of mobility as an aesthetic character. In addition to this he mentions the familiar coldness, and the feeling of emptiness and sadness. The important thing to remember about these comments is that Goethe talks here of blue as a total color, so to speak. In section 763 he recommends staying in a monochromatic room or looking at the world through a correspondingly colored glass in order to study this effect. "Then one identifies oneself with the color; it draws both the eye and the spirit in unison." The other well-known effects of blue appear later under the key words of harmonic and characteristic combinations as well as combinations without character. We are now faced with the question: What has blue got to do with the experience of cold, with the feeling of empty space, how does blue contain the suggestion of movement, and what is it about blue that makes us melancholic? Goethe seems to have wanted to give an explanation for this characteristic trait of blue by recalling his genetic derivation of the color blue (782), i.e. that blue always contains something dark (778). According to Goethe blue arises in darkness or blackness conditioned by gloom. However, this explanation could only be relevant for the synesthetic character of blue to distinguish it from other colors, and not to show that blue and other colors have such characteristics per se. To do this one would have once again to pose the principal question as to what it in fact means to see blue, or in more general terms, as to what perception is in fact.

Theories of synesthesia

Theories of *synesthesia* usually suffer from the fact that the phenomenon of perception is approached from the standpoint of natural science. Seen from this point of view, the first given consists of physical stimuli and the other consists of the sensory organs. This approach erects seemingly insurmountable barriers between tones, color, smell, and the quality of touch. What had already been categorically different from the point of view of physical stimulus has now been further separated by the specific reception through the respective sensory organ. The discovery of the so-called specific sensory energies seemed to teach us that even in cases in which stimuli were identical, they were always perceived specifically by the various sensory organs. Pressure on the eye creates the impression of light, while pressure on the skin creates the feeling of pressure. Against the background of such an approach *synesthesia* in themselves became impossible or rather they demanded special theories extending beyond the domain of perception. The situation was aggravated by the philosophical backgrounds to such a view of perception. From Locke to Kant to Ernst Mach, the primacy of sensation remained predominant in philosophical epistemology. These philosophers conceived of the epistemological process as a process in which we were given unrelated material through the senses in the form of sensations, and that all unity in the perception of an object came from the subject, or more specifically from the rational mind. According to these conceptions *synesthesia* ought to be products of the subject, i.e. associations, otherwise one would be dependent on purely empirical contiguities (the contiguity thesis). An attempt to save this would be to appeal to a so-called *sensus communis*. The doctrine of *sensus communis* has been erroneously traced back to Aristotle, as evidence has shown.[3] However, in his treatise on the soul, Aristotle raises the question as to how one could distinguish, for instance, between sweetness and light and – we believe – establish a relationship between sweetness and lightness. At the time Aristotle solved this problem by a tour de force in claiming that all the senses were interconnected like lines to a point and were thus – in certain way – one. Later on the doctrine of *sensus communis* was developed from this, i.e. the assumption of a sense which existed over and above the five senses, in which the individual sensory domains converged. This may, however, provide a clue to the solution of our problem, but the assumption of an additional sense will not help at all. For, in that case, the question of *synesthesia* existing between this sixth sense and the other senses would arise. Wilhelm Wundt's solution foundered on precisely these problems. Wundt claimed that *synesthesia* derive from the relationship between the different senses and "feelings." Thus for example, to remain with the example cited from Goethe above, melancholia would be a synesthetic character which blue shared with deep, soft notes, for instance, and it would have its basis in the related effect of these two sensual qualities on the senses. In this connection Hermann Schmitz[4] rightly pointed out that feelings also have a synesthetic character so that, as *tertium comperationis*, they could not be considered in relationship to the five senses. Rather one would somehow have to find a *quartum comperationis* alongside the senses and feelings. Schmitz demonstrates this by pointing out that colors, like melancholia, are able to share the synesthetic character of darkness, for example.

The psychological theories of *synesthesia* have now gradually dissolved in the face of empirical research. It has, for example, been shown that two phenomena from different domains of the senses cannot be synesthetically linked by frequent simultaneous appearance. The continuity crisis stops there. Furthermore, it has been shown that the frequency of synesthetic experiences declines with age. This speaks against the supposition that they might be biographical associations. Heinz Werner[5] in his reference book on psychology summarizing developments up to 1966 notes: "One can say that explanation by association ... has generally lost ground" (290). The observation that *synesthesia* play a greater role among children and primitive peoples than among "civilized peoples" (288) leads him to conclude that *synesthesia* belong to a deeper level than perception.

> Interestingly enough, one can ... uncover layers among civilised people which, genetically, exist prior to perception and are partially buried as original modes of experience in the "objective" type of person. Environmental stimuli do not appear in consciousness as objective perceptions in this layer, but as expressive sensations feeling the entire ego. In this layer it is indeed the case that notes and colours are "experienced" rather than perceived.
>
> (297)

The term "sensations" used, rather unfortunately perhaps, here could be understood as a relapse to Wundt's theory. But this is not the case. Werner uses this term to refer to the domain of physical experiences. The deeper layers of which he speaks are the more or less noticeable states of the body. For this reason, he summarizes his conceptions as follows:

> It is highly likely that the unitary synesthetic experience, which may be triggered in the realm of vital experiences by the most divers stimulus modalities, is caused by movements of tension in the body.
>
> (298)

The philosophical theory of synesthetic character advanced by Hermann Schmitz follows on directly from this work, which is still probably the most progressive position in the psychology of the senses.[6] In contrast to Werner,

Schmitz is in the fortunate position of being able to elaborate his conception of *synesthesia* from an existing and developed theory of human corporeality. He does not, therefore, talk of "movements of tensions in the body," but of bodily sensations. He shows that the phenomena generally termed *synesthesia* are basically the character of feelings. Of decisive importance here is the fact that the synesthetic character as such, i.e. without external sensual perception, can also be felt sensuously. Thus, he points out, for example, that heaviness, which can be attributed to a sound owing to its synesthetic character and can be felt, as such, physically in the case of tiredness or stupefaction.[7] The fact, demonstrated by Schmitz, that feelings like sadness or anger can share certain synesthetic characteristics with the sensual qualities may be explained by the fact that feelings intervene in the economy of the body, i.e. in its relations of tension, as Werner would say, and can hence be experienced physically.

Thus, according to Schmitz, synesthetic characters are experiences of a certain kind, i.e. characteristics of one's physically feelings. The avoidance of the concept of *synesthesia* can be explained in this manner: for it is not a question of a quality which belongs to one sensory domain – such as lightness in the domain of optics – being ascribed to phenomena of a different realm of the senses – such as that of music – in a manner not intrinsic to its character. Although it is possible for a synesthetic character to be described in the terminology of another sensual domain, in which it primarily appears, the decisive thing, however, is that the so-called *synesthesia* arise because sensual perceptions definitely reach down into the realm of bodily feelings. Thus, Aristotle's punctual unity of the senses now appears to be extended to constitute a *sensus communis*. Does this mean that feeling something physically is an additional sense on top of the other senses? Following philosophical tradition one is tempted to talk of an inner sense here. However, the inner sense was considered to be reflexive, as in the case of Kant, for example: through the inner sense the mind becomes aware of its own condition. However, this is out of the question. Neither the tension changes of the body, of which Werner speaks, nor the physical feelings of Hermann Schmitz are in any way reflexive. Both Werner and Schmitz avoid all talk of another sense. For Werner the tension changes are phenomena accompanying perception, but they are not themselves perception. Schmitz, by way of contrast, declares that the experience of synesthetic characters (alongside that of movement suggestion) to be the real and fundamental characteristic of perception. For him, perception is basically corporeal communication (*Wahrnehmung*, p. 69). That is, however, a radical thesis. It leads Schmitz, counter to his other intentions, to a kind of projection thesis. The appearance of *synesthesia* in the various realms of the senses is referred to as "reflection" (*Subjektivität*, p. 63) or "radiance" (*Subjektivität*, p. 52). The synesthetic characters which are thus in fact supposed to belong to the bodily feelings are invested, by Schmitz, in the external sensory data. As a consequence of this radicality, the phenomena introduced above in the quotes from Goethe are in danger of becoming lost: for instance, that a blue can be called cold because coldness belongs to its reality (its energy as a color).[8] Let us raise the question once again – now that we have gathered our strength a little analyzing Hermann Schmitz – as to the nature of the perception of blue.

What is perception?

This century perceptual psychology has gradually recovered perception in all its splendor from the thesis of the primacy of sensation. It is not a question of individual sensual data, which one could synthesize to surfaces, figures and things, but entire surfaces and forms. Indeed, from the very start, one does not only see forms but things. But that is not true either. One sees things in an arrangement, things which exist in relationship to one another and one sees situations. The philosophy of gestalt psychology adds that these are also already embedded in particular totalities. Each specific situation can only become concrete against the background of a world. Yet one doesn't see the world. So what is this totality in which all particulars are embedded on which one is able to focus, depending on the degree of attentiveness and analysis? We call this primary and, to a certain degree, basic object of perception the atmosphere. Its priority becomes apparent with a switch of perception or – to be more direct, taking an analogy from film techniques – with a cut in which one enters another world, as it were. For example, one leaves a lively street and enters a church. Or one enters a flat that one is not yet familiar with. Or one stops for a rest during a car trip, takes a few steps and suddenly has a view of the sea. In such primary situations it becomes clear what is perceived first and, above all, in detail; in a certain sense it is space itself. Space is not taken to mean the Kantian sense of pure observation of things existing external and adjacent to one another, but the emotionally influenced restrictions or expansiveness which one enters, the overwhelming fluidum. We call this the atmosphere, borrowing our terminology from Hermann Schmitz. One enters a flat and is overwhelmed by the philistine atmosphere. One enters a church and has the feeling of being shrouded in a holy gloom. One catches sight of the sea and is swept off into the distance. It is only against this background or in this atmosphere that one can distinguish the details. One can recognize things, name colors, identify smells. The important thing is that each individual aspect is in some way colored by the atmosphere. The furniture imposes itself in the cramped philistine space; the blue of the sky seems to be in flight, the empty benches of the church invite one to be devout. This, at least, is how the person perceiving experiences the situation. The aesthetic worker knows differently. He knows how to create atmospheres by designing rooms, by using colors and requisites.

How does one perceive atmospheres? One thing is certain and that is the decisive thing for us here: it is not by one single sense or by their interplay. "I see this, I hear this, I smell this" are processes which occur in the second stage, already constituting the beginning of the analysis. In colloquial language one would probably answer the question as to how one perceives atmospheres as follows: "By intuition. I just feel it." But that would be going too far, understanding – together with Hermann Schmitz – this feeling as a corporeal feeling, and as we have seen, making the synesthetic characteristics as characters of physical sensations. Physical sensations already contain too much reflection, being a sensing of oneself, whereas perception of the atmosphere means quite simply that the atmosphere is the thing which is perceived. One could express this as follows: in perceiving the atmosphere I feel the nature of the environment around me. This perception has two sides to it: on the one hand the environment, which "radiates" a quality of mood and, on the other hand, me participating in this mood with my sensitivity and assuring myself that I am here. Perception qua sensitivity is presence which can be sensed. In turn atmospheres are the manner in which things and environments "present" themselves.

To return to *synesthesia*: It can be seen that the primary and basic phenomenon of perception, i.e. the atmosphere, does not have any character based on individual senses at all. If one wanted to follow the pseudo-Aristotelian tradition of *sensus communis* one would have to say that sensitivity is the *sensus communis*. Schmitz is right in stating that the synesthetic characters (alongside movement suggestions, which we have not considered here) are of more fundamental importance for perception than "the supposed acts or sensations of seeing, hearing etc." (*Die Wahrnehmung*, p. 69).

The fact that blue is experienced as cold or a note as sharp results from its analytic origins in specific atmospheres. Conversely, from the standpoint of the aesthetic worker one could say that a blue "works atmospherically" and radiates a specific atmosphere. The perception of a blue is only to a very small extent, and in the last analysis, the authority demonstrating that the color blue exists at a certain place. Even a blue which may be localized has a radiance, as every painter knows, effecting its environment. And it is not only perceived at the place where it actually is, but in the entire space, somehow. In order to study this atmospheric effect of color, Goethe rightly recommended exposure to a room which was completely blue, and avoiding the objectifying mode of perception – the objective attitude of a civilized person, as Heinz Werner said. The perception of blue then ceases to be the recognition of this color but sensing the atmosphere, i.e. how I feel in this atmosphere. Then one can see that coldness and, as Goethe said, emptiness are indivisibly linked up with this blue.

The production of atmospheres in architecture

One could raise the objection to what has been stated above that the "objective attitude of the civilized person" in our technical civilization always hides the "deeper layers" of perception and primarily perceives things, or maybe not even things but signals. Indeed, what one perceives is very dependent on the socialization of perception and on the respective situation in which one has to act. Nevertheless, the ability to sense atmospheres is never lost. Even though it may not enter into consciousness it still has an effect on the way one feels.

Architects have to reckon with this, and do so in fact. Architecture in particular produces atmospheres in everything it creates. It does, of course, solve objective problems and build objects, buildings of all descriptions. But architecture is aesthetic work inasmuch as rooms and space are always created with a specific quality of mood and hence as atmospheres. Buildings, interior rooms, squares, shopping centers, airports, and urban spaces such as cultural landscapes can be elevating, oppressive, light, cold, comfortable, solemn, and objective; they can radiate a repelling or an inviting, an authoritative, or a familiar atmosphere. The visitor and user, the customer and the patient are all touched or moved by these atmospheres. The architect, however, creates them, more or less consciously. The sensual items which he posits: the colors, the design of surfaces, the lines, the arrangements and the constellations are, at the same time a physiognomy from which the atmosphere emanates. This is a matter of course to every architect as an aesthetic and as a practical worker. The sensual properties which he gives to his products are less relevant as such, being relevant primarily in the wealth of their synesthetic effects. Nevertheless, his consciousness is generally directed toward the properties and determinations he gives his products. What the philosopher had to recall was the notion that it is never purely a question of designing an object but always, at the same time, of creating the conditions for its appearance.

Notes

1 As for the introduction of the concept, see my book *Für eine ökologische Natur-ästhetik*, Frankfurt/M., Suhrkamp, 1989 and the interview with the same title guided by Florian Rötzer, in Fl. Rötzer (ed.), *Digitaler Schein*, Frankfurt/M., Suhrkamp, 1991.
2 G. Böhme, "Ist Goethes Farbenlehre Wissenschaft?," in G. Bohme (ed.), *Alternativen der Wissenschaft*, Frankfurt/M., Suhrkamp, 1980.
3 W. Welsch, *Aisthesis*, Stuttgart, Klett-Cotta, 1987.
4 Hermann Schmitz, *Subjektivität*, Bonn, Bouvier, 1968, "Gemeinsinnliche und einzelsinnliche Qualitäten," S. 67.
5 H. Werner, "Intermodale Qualitäten (Synästhesien)," 9. Kap., *in Handbuch der Psychologie*, 1. Bd., 1. Halbband, Göttingen, 1966.
6 In addition to the already mentioned work of Hermann Schmitz, see his *System der Philosophie*, Vol. III, 5, Die Wahrnehmung, Bonn, Bouvier, 1978.
7 Schmitz, *Subjectivität*, p. 62.
8 For the introduction of the concept *Atmosphere* see in particular Hermann Schmitz, *System der Philosophie*, Vol. III, 2; "Der Gefühlsraum," § 149ff., Bonn, Bouvier, 1969.

7   Contribution to the critique of the aesthetic economy[1]

Introduction

With their essay on the culture industry in the *Dialectic of Enlightenment*, Horkheimer and Adorno established a paradigm which, from the 1940s to today, has provided the indispensable reference point for every critique of aesthetic production in its relationship to the economy. This is not to say, however, that the most urgent task today in the debate around the culture industry theory would be to prove its continuing productiveness, nor would it be, conversely, to underline the theory's shortcomings – that it fails to differentiate between popular and mass art, and thus does not recognize the possibilities of subversive pop art; that its distinction between high and low culture feeds off the self-consciousness of an educated elite; that it disparagingly characterizes cultural consumption as an illusory satisfaction.[2] Rather it seems necessary to me today, after the almost total retreat of class-specific consciousness, after a radical transformation of the economic significance of aesthetic production, and after the transition since the 1950s into a new phase of capitalism, to pay our respects to the culture industry theory by reconstructing it in the light of these changed conditions. I recommended such a reconstruction several years ago in my "Sketch of an Aesthetic Economy,"[3] and I will begin here by recapitulating that sketch in condensed form.

The aesthetic economy starts out from the ubiquitous phenomenon of an aestheticization of the real, and takes seriously the fact that this aestheticization represents an important factor in the economy of advanced capitalist societies. The concept of aesthetic labor must first be developed if this situation is to be grasped. Aesthetic labor designates the totality of those activities which aim to give an appearance to things and people, cities and landscapes, to endow them with an aura, to lend them an atmosphere, or to generate an atmosphere in ensembles. With this term, the qualitative valuation of the products of aesthetic labor, and with it the distinction, so essential to the culture industry theory, between art and kitsch, is quite consciously abandoned. The assumption that a gulf separates the creators of art or culture from artisans, cosmeticians, and advertisers is also left behind. The concept of the aesthetic laborer encompasses rather the entire spectrum from painter to artist, from designer to music producer; it embraces all human activities that lend to things, people and ensembles that *more* which goes beyond their handiness and objective presence, their materiality and practicality. Because this *more* has attained its own economic significance, the concept of staging value (*Insze-nierungswert*) was coined. With this term, the Marxist dichotomy of use and exchange value was expanded to include a third value category. The use value of a commodity consists in its practicality within a determinate context of use. The exchange value of a commodity consists in the value it is accorded within the context of an exchange process, and is abstractly measured in money. In order to raise their exchange value, however, commodities are treated in a special way: they are given an appearance; they are aestheticized and *staged* in the sphere of exchange. These aesthetic qualities of the commodity then develop into an autonomous value, because they play a role for the customer not just in the context of exchange but also in that of use. They are certainly not classical use values, for they have nothing to do with utility and purpo-siveness, but they form, as it were, a new type of use value, which derives from their exchange value insofar as use is made of their attractiveness, their aura, their atmosphere. They serve to stage, costume, and intensify life.

It is decisive for the aesthetic economy that a quantitatively significant sector of the national economy be geared to the production of values for staging and display, or rather that giving commodities a staging value makes up an essential part of their production. As such, the values produced by the aesthetic economy are to a large extent not actually needed. The aesthetic economy thereby proves to be a particular stage of development of capitalism. Capitalism, along with every other economic form, is usually regarded as an organization for overcoming human needs and satisfying material requirements. There have certainly always been very clear-sighted theorists, such as Veblen[4] or Sombart,[5] who explicitly connected capitalism with luxury production, and their books must be judged today as the beginnings of a critique of the aesthetic economy. But, on the whole, capitalism was and is seen as a highly serious business, to be assigned, borrowing Marx's terminology, to the realm of necessity, not to the realm of freedom. At a certain stage of development in which the material needs of society are generally satisfied, capitalism must bet upon another type of needs, which calls for the appropriate term *desires*. The third fundamental category of the aesthetic economy is thereby named. Desires are those needs which, far from being allayed by their satisfaction, are only intensified. Needs in the narrower sense, for example the need to drink, to sleep, and to find shelter from the cold, vanish the moment they are stilled. Desires are quite different: the powerful want ever more power, the famous still more fame, and so on. It is important to note that there are desires that can be directly commercially exploited, namely those that are directed toward the staging, and hence the intensification, of life. There are no natural limits to presentation, glamour, and visibility. Each level, once reached, demands instead its further intensifica-tion. Because growth belongs intrinsically to capitalism, when capitalist pro-duction attains a particular stage of development typified by the fundamental satisfaction of people's material requirements, it must henceforth explicitly turn to their desires. The economy thus becomes an aesthetic economy.

Reconstruction of critical theory

The critique of the aesthetic economy should make a contribution to the reconstruction of critical theory. By reconstruction is not to be understood the philological appropriation of critical theory but its historical continuation. It is not just a question of analyzing changes in basic social, economic, and technological conditions but also of reexamining successive theories on the relationship between aesthetics and the economy in the light of those changes. As far as the culture industry essay is concerned, one must take a further step back to Walter Benjamin, who in his essay "The Work of Art in the Age of Mechanical Reproduction"[6] had already thematized artistic genres – photography and film, at the time – for which mechanical reproducibility is essential. Moreover, with the appearance of these artistic genres, he focused on the emergence of a specifically mass culture. Horkheimer and Adorno's essay must be described as regressive in this respect. They equate mechanization with standardization, seeing in the industrial production of art nothing other than its degradation.[7] Both were familiar with Benjamin's essay. They also knew a few of the key ideas from his unfinished *Arcades Project*, including the idea that the aestheticization of the commodity world – a process whose inception, as is well known, Benjamin already locates in the nineteenth century – is a basic feature of high capitalism. With this Benjaminian concept the perspective of a culture industry theory is significantly broadened. For when Adorno and Horkheimer speak of art, they are thinking primarily and perhaps even exclusively about artistic production and reception and their critique is concerned with art's commercialization and trivialization. There are of course thoughts in the culture industry essay in which the broadening of aesthetics made possible by Benjamin's observations on the aestheticization of the commodity world already leaves its mark. On the one hand, Horkheimer and Adorno see in the culture industry by no means just an annexation of art from without, but also a development arising out of art's own logic.[8] On the other, they also glimpse in culture's decline into industrially produced entertainment the chance for the latter's aesthetic revaluation:

> The fusion of culture and entertainment that is taking place today leads not only to a depravation of culture, but inevitably to an intellectualization of amusement.[9]

One step further and they would have realized that the avant-gardist demand for the transformation of art into life could also succeed: that art, once it had submitted to industrial production and been appropriated by advertising, could reappear, transformed, in design. The qualitative differences of which Horkheimer and Adorno make so much, and whose leveling they greet with a stream of invective – trash and swindle – would then no longer mark territorial boundaries between, say, art and arts and crafts, but could be applied to all sectors of aesthetic labor. This expansion of the theme of culture to aesthetic labor as a whole would necessarily shift the fronts of critique. In future, they would no longer be dictated by the polar opposition of art and its industrial depravation.

Thus transformed, Horkheimer and Adorno's theme was taken up again by Wolfgang Fritz Haug and Jean Baudrillard at the time of the European economic miracle. While Haug's *Critique of Commodity Aesthetics*[10] no longer investigated the relationship of industrially produced aesthetic works to art, it was still concerned, like the culture industry essay, with the repressive character of the aesthetic economy. The culture industry essay had already made mention of the film's homogenizing power over feeling and life form,[11] and of the denial perpetrated by the culture industry through the illusory satisfaction it offers. "The culture industry perpetually cheats its customers of what it perpetually promises."[12] Haug very vividly condenses this thought by describing the aestheticization of commodities in terms of their sexualization. He sees in this process an education to voyeurism.

> In a situation of general sexual repression, or at least of isolation, the use-value of mere sexual illusion lies in the satisfaction which voyeurism can provide. This satisfaction through a use-value, whose specific nature is an illusion, can be called illusory satisfaction. The characteristic of this satisfaction through sexual illusion is that it simultaneously reproduces further demand alongside satisfaction, and produces a compulsive fixation.[13]

This passage reveals as much the far-sighted gaze of the author as its restriction through his ascetic marxism. Pleasure in the aestheticization of the real, disqualified by Horkheimer and Adorno as "amusement." proves an illusory satisfaction for Haug as well – and it lies outside of Haug's horizon that the Eros of distance, which according to Klages is kindled by the "reality of images." might be the true Eros. He proves equally incapable of developing the thought announced at the end of the quotation: that a satisfaction might simultaneously reproduce further demand. That is only possible if, following Bataille, we distinguish desires from needs and characterize capitalism as an economics of luxury expenditure. Haug, meanwhile, continues to insist that the economy serves the satisfaction of needs. He defends authenticity against semblance, the commodity's use value against its aestheticization.

The critique of commodity aesthetics is thus, more precisely, a critique of the aestheticization of the commodity world. Haug cannot or will not, therefore, concede an autonomous status to the aesthetic value of commodities. For him, commodity aesthetics is the more or less successful mediation of a contradiction which erupts in the commodity character itself: the contradiction between exchange and use value. To be able to realize the exchange value of the commodity in the act of exchange, the seller must invest it with qualities that make it attractive to the customer – who is actually only interested in its use value – *in* the act of exchange as well:

> Henceforth in all commodity production a double reality is produced: first the use-value, second, and more importantly, the *appearance* of use-value.[14]

For Haug, this second reality, the actual aesthetic character of the commodity, appeals to the customer's need for a use value. Haug thus discerns the promise of a use value in the aestheticization of the commodity.[15] It is difficult to judge whether the commodity aesthetic had yet to become an independent value in the phase of economic development when Haug wrote these lines (i.e. around 1970), or whether his dialectical derivation caused it to escape his attention. In any event, the commodity's aesthetic value appears to him not in its conceptual autonomy, but only as a material independence – as packaging. The aestheticization of the commodity leads to a sort of aesthetic commodity couture, worn only in the context of exchange and cast aside in the context of use.[16]

Haug's critique of commodity aesthetics can thus be summarized as follows: a promise of use value is made by the producer or seller that, in the end, cannot be redeemed by the commodity. Haug fails to see that the commodity aesthetic satisfies a need of the customer which is *not* aimed at use value.

Things are quite different in the works of Jean Baudrillard that appeared around the same time. Of these, *La société de consommation, ses mythes, ses structures*[17] is predominantly descriptive, while *Pour une critique de l'économie politique du signe*[18] endeavors to provide a theory of "consumer society." Baudrillard's catchword already heralds the decisive turn. Consumers are not simply concerned with satisfying their needs through the purchase of use values. Rather, they are after consumption as such, because through consumption they can make manifest their social position. Baudrillard attempts to clarify this difference, which he takes from Thorstein Veblen,[19] by recourse to the symbolic exchange described by Malinowski, the kula, or potlatch. Whereas commodity exchange usually involves the satisfaction of primary needs, those who take part in the kula lavish extravagant gifts upon each other in order to acquire or manifest social status.[20] Baudrillard claims that this difference between the two forms of exchange vanishes in consumer society. What is more, symbolic exchange comes to dominate customary commodity exchange:

> The fundamental conceptual hypothesis for a sociological analysis of "consumption" is *not* use value, the relation to needs, but *symbolic exchange* value, the value of social prestation [sic], of rivalry and, at the limit, of class discriminants.[21]

The interest that the customer brings to the commodity is thus, according to Baudrillard's analysis, the interest in buying it in the first place, in being able to afford it at all. Baudrillard recognized that the commodity character – or better put, the appearance of things as commodities – continues to exist outside the exchange process. The aestheticization of commodities, for Haug solely mediation between use and exchange value, thereby wins its autonomy. It is no longer a mere mediation, but a use value of the exchange value, so to speak. Baudrillard can therefore name a third value type alongside use and exchange value, which he calls "sign value" (*valeur signe*). This occurs explicitly in the chapter "Toward a Critique of the Political Economy of the Sign." Yet upon closer inspection, one sees that Baudrillard is still not willing to concede a value in the context of *use* to the commodity's "form of appearance" in exchange, i.e. its aesthetic. In his effort to prove the economy of consumer society to be an amalgam of commodity exchange and symbolic exchange, he attempts instead to construct a parallel between exchange value and use value on the side of the sign. For Baudrillard, the sign's signifier corresponds to exchange value, its signified to use value.[22] Certainly, this allows him to determine the sign-character of commodities and its importance for the discrimination of customers – by wearing a particular fashion accessory one is marked as fashion-conscious, by purchasing a particular piece of china as a person of taste – but through this abstraction he loses the link to the aestheticization of the commodity world, which is what concerns us here. For the sign-character of the commodity in the context of exchange, which gives the customer social standing, can only be in the *price* of the commodity. Baudrillard's economics of the sign is thus, at heart, a theory of the status symbol, and his critique of consumer society that of a society stratified according to levels of consumption. Nonetheless, the importance of Baudrillard's approach should not be underestimated. Books like Bourdieu's *Distinction* and Gerhard Schulze's *Die Erlebnisgesellschaft* doubtless profited from his advance.

Another thing is important: in passing from an economy of needs and use values to an economy of signs, Baudrillard implicitly moves from one of scarcity and ascesis to one of abundance and luxury expenditure. This becomes particularly clear in comparison with the contemporaneous works of Wolfgang Fritz Haug. I have already mentioned Baudrillard's debt to Veblen, and hence to a socioeconomic theory which sees consumption as serving the realization of social status, rather than the satisfaction of primary needs. He could just as well have sought inspiration in the works of Werner Sombart, who downright identifies the origin of capitalist economics in luxury and profligate expenditure. His book *Luxury and Capitalism* showed that luxury and excessive consumption draw especially upon aesthetic values – in the ostentatious display of life through clothing, colors, perfumes, and mirrors, or, in everyday consumption, in the refinement of food through spices and its supplementation through semi-luxury goods like coffee, tea, and sugar. His analyses are of extraordinary importance, for they foreground the insight that the economy of profligate expenditure is by no means simply a matter of quantitatively increased consumption and the elimination of surpluses, but rather of a different type of consumption and a different type of needs, which I have termed desires. Desires are intensified through their satisfaction; they serve life's intensification in general, not its preservation. Terminologically, this difference could probably be better expressed in French, through the distinction between *besoin* and *désir*. In this regard, it is the great merit of Georges Bataille to have brought to light the essence of another economy, an economy of excessive expenditure. Bataille went so far as to call it "the sublation of economy," although he also understood it to be only one side of general economy, which moves dynamically between phases of accumulation and phases of prodig- ality. Coming after Veblen, Sombart and also Schumpeter,[23] Bataille made no startling contributions to the tradition of economics. Yet it must be emphasized that, for Sombart as for Veblen, profligate expenditure was always only the prerogative of the ruling classes in a society which could be split into producers and consumers. As such, only once the trickle-down effect already sketched by Sombart and Veblen – the extension of the luxury economy to *all* of society – had become apparent could the anthropological depth of Bataille's works find widespread recognition. Only in such a situation, known to us as the affluent society, did it become obvious that while luxury and excess may contribute to the distinction and realization of social status, they are also something more generally human, and therefore relate to a fundamental human need. This need, which is not a primary need (for it does not serve the preservation of life), is the wish for life's intensification.

Such a social and economic condition was probably reached in Europe and North America in the 1950s and 1960s. It is the condition in which capitalism, having in principle satisfied primary needs, must turn to desires. It is the phase in which a large part of social production becomes aesthetic production, serving staging values instead of use values.

The completion of this transition required a cultural revolution, entailing the rejection of the Protestant ethic, the overcoming of ascetic values and the rehabilitation of desires. This cultural revolution came about with the student movement, or more precisely with a particular faction of the student move- ment.[24] Its literary expression is to be found, not in the ascetic Marxism of a Haug, but in Bataille, and above all in Marcuse's *Eros and Civilization* (1955).

Marcuse's theme is not economics but culture, so the author he sets to work upon is not Max Weber but another theorist of ascetic morality, Sigmund Freud. Against Freud, he attempts to show that culture need not be repressive. Freud had sought to justify the renunciation demanded of the individual by appealing to the reality principle, which prevents the individual living his or her life to the full in the interest of collective survival. Marcuse, in contrast, differentiates between reality principle and performance principle; specifically, he shows that the former is actually an instance of the latter. Freud's reality principle compels more renunciation than would be necessary for coping with reality and staying alive, and this extra renunciation serves the maintenance of domination. In other words, more renunciation is demanded of the individual than is absolutely necessary. Marcuse's perspectival shift makes visible an economic situation in which primary needs are satisfied and the pleasure principle can be accommodated. At the time of writing the book, Marcuse seemed uncertain as to whether this situation had already been achieved. On the one hand, he sees clearly enough that the American standard of living rested upon international dependencies;[25] on the other, he still believes that "the reduction of the working day ... would almost certainly mean a considerable decrease in the standard of living prevalent today in the most advanced industrial countries."[26] He therefore writes cautiously: "The reconciliation between pleasure and reality principle does not depend on the existence of abundance for all."[27] As it happened, the transition into the affluent society in the 1960s provided the factual basis for Marcuse's ideas becoming the ideas of a cultural revolution. It is scarcely possible today to gauge the extent to which they transformed lifestyles and ways of thinking in the advanced industrialized nations. The liberalization of sexual life, the rehabilitation of pleasure, the development of leisure culture, the expansion of the ludic element in life, and the gradual obsolescence of the idea that work is life's fulfillment number among these changes. They include, further, the transfer of interest from reality to its appearance; virtual realities and the *mise-en-scène* of cities; aesthetics of existence and the ethics of the good life.

Marcuse had oriented his utopian reconciliation of reality and pleasure principles on Schiller's *Letters on the Aesthetic Education of Humanity*: the realm of freedom as the realm of play. It begins beyond the socially necessary minimum of repressive labor. Of course, he does not overlook – and here we return to Horkheimer and Adorno's "culture industry" – that even leisure time could once again be placed under the performance principle:

> Not until the late stage of industrial civilization ... has the technique of mass manipulation developed an entertainment industry which directly controls leisure time.[28]

Shifting the fronts of critique

The line of development in the critique of the aesthetic economy that stretches from the "Culture Industry" essay through *Commodity Aesthetics* to *Eros and Civilization* has already made visible a shift of the fronts of critique. At the same time, its field, like the field of aesthetics overall, has been significantly expanded, from art through commodity aesthetics to the aestheticization of life as a whole. A revision of the enterprise paradigmatically presented by Horkheimer and Adorno in their "Culture Industry" text must today be articulated, first and foremost, through a renewed determination of the critical front.

Three basic conditions must be respected. All three are closely connected, and today they let aesthetic production as well as aesthetic consumption appear in a different light. The first basic condition is the state of capitalist development itself. One can indicate this state with the catchwords "consumer society," "affluent society," or "luxury economy." And one can regard it, from a global perspective, as a local phenomenon: only in those few societies marked by this phase of capitalism is a large part of production and consumption determined by aesthetic values. The second basic condition is a transformed attitude toward the pleasure principle: the good life is no longer determined through work, saving, and ascesis, but through leisure, consumption, and play. The third basic condition is the end of class society. This condition, which has not previously been considered, requires a brief elucidation.

The line of aesthetic consumption drawn by Baudrillard and Bourdieu, namely from the economy of signs to that of distinctions, seems to imply that class differences are reproduced in aesthetic consumption. Yet that is an illusion, insofar as the classes described by Bourdieu are not the traditional ones, which reproduce themselves through their position vis-à-vis the forces of production, and therefore represent a structure of domination, but – as he himself writes – "constructed classes,"[29] i.e. sociological *categories* formed through the clustering and inner coherence of a mass of distinguishing features. Bourdieu still speaks freely of ruling classes, of upper and lower classes, but what really matters to Bourdieu is distinction as such, differentiation from other social groupings. The economy of signs, which as late as the 1960s may still have spawned a hierarchy of status symbols, has yielded today to a signaling and staging of group affiliations, articulating a multiplicity of group styles and life forms that have little to do with social stratification and class domination.

Under these conditions, a revision of the critical enterprise begun by Horkheimer and Adorno can no longer target art's depravation. It can be credited to their middle-class background that their critique drew from the source of true, authentic, high art. But today we must realize that neither the commodity character of it nor its technological–industrial reproducibility have done it damage, and that so-called mass culture is neither blandly uniform nor repressive per se. So far, an obdurate avant-garde has always been able to assert itself, while the quality standards of aesthetic production, far from entering a terminal decline, have proved capable of developing in mass culture as well. And out of that culture, subversive movements have been able to emerge, movements which stand at cross-purposes to dominant social and work practices.

One must further hold against Horkheimer, Adorno, and also Haug that aesthetic consumption, from film and television through advertising to commodity aesthetics, can by no means be disqualified as illusory satisfaction, as quid pro quo, or even as an additional deception of the public. This type of critique relies upon the traditional difference between being and appearance, reality and mere semblance. Instead, we have to realize that the advent of virtual reality has established a new area of life situated beyond labor regimes and life's grim seriousness, an area in which people today can invest their emotions and playfully and pleasurably rehearse desired life forms. To condemn it would, in the wake of Marcuse, be pure prudishness. Against Marcuse, however, one must recognize that the gates to the realm of freedom are not thereby flung open. And so the old critique steps forth in a new guise.

For the aestheticization of reality driven forward by the aesthetic economy is thoroughly ambivalent. One must first ascertain that the economy of excessive expenditure, rehabilitated by Bataille, and with good reason, against Puritanism and Victorianism, still depends on worldwide exploitation. Further, the establishment of a world of images on the surface of reality, or even independently of it, may well serve the intensification of life, but it should not make us forget that society is neither just imaginary (Castoriadis), nor does it occur as a play of simulacra (Baudrillard), but ultimately still rests on relations of violence. And, finally, one must ask with Horkheimer and Adorno, and both with and against Marcuse, why the aesthetic economy does not open the gates to the realm of freedom. Under these changed conditions, why does Horkheimer and Adorno's dictum still hold true: "Amusement under late capitalism is the Prolongation of work"?[30] And on what does that "technique of mass manipulation" rest which, according to Marcuse, "develop[s] an entertainment industry which directly controls leisure time"?[31] It rests, we can answer, upon the fact that the aesthetic economy must necessarily bet upon desires, i.e. upon needs which are intensified rather than allayed by their satisfaction. The development of these desires – desires to be seen, to dress up, to stage oneself – forms the basis for a new, practically limitless exploitation. On this basis, consumption can become an obligation, affluence a stress, extravagance a duty. Under these conditions, one will probably have to define the sovereign human being[32] differently to Bataille, and expressions like "solidarity," "seriousness," and "ascesis" could acquire an altogether new meaning.

Notes

1 Translation by Robert Savage.
2 For this type of critical retrospective, see *Zeitschrift für kritische Theorie*, 10, 2000.
3 First printed in my essay *Atmospäre als Grundbegriff einer neuen Ästhetik*, in *Kunstforum International*, 120, 1992, 237–55. Now in my lectures on aesthetics, published as *Aisthetik. Vorlesungen über Ästhetik als allgemeine Wahrnehmungslehre*, Munich, Fink, 2001.
4 T. Veblen, *The Theory of the Leisure Class*, London, Allen & Unwin, 1970 [1899].
5 W. Sombart, *Luxury and Capitalism*, trans. W. R. Dittmar, intro. P. Siegelman, Ann Arbor, University of Michigan Press, 1967, [1913].
6 W. Benjamin, "The Work of Art in the Age of Mechanical Reproduction," in Hannah Arendt (ed.), *Illumination*, trans. Harry Zohn, London, Fontana, 1992 [1939].
7 "Movies and radio no longer pretend to be art. The truth that they are just business is made into an ideology in order to justify the rubbish they deliberately produce." "A technological rationale is the rationale of domination itself. It is the coercive nature of society alienated from itself ... It has made the technology of the culture industry no more than the achievement of standardization and mass production, sacrificing whatever involved a distinction between the logic of the work and that of the social System" (M. Horkheimer and T. Adorno, *Dialectic of Enlightenment*, trans. J. Cumming, London, Allan Lane, 1973 [1947], p. 121).
8 "Having ceased to be anything but style, it reveals the latter's secret; obedience to the social hierarchy. Today aesthetic barbarity completes what has threatened the

creations of the spirit since they were gathered together as culture and neutralized. To speak of culture was always contrary to culture" (ibid.).

9 Ibid., p. 143.
10 W. F. Haug, *Critique of Commodity Aesthetics*, trans. R. Bock, Cambridge, Polity Press, 1986 [1971].
11 "Real life is becoming indistinguishable from the movies" (Horkheimer and Adorno, *Dialectic of Enlightenment*, p. 126).
12 Ibid., p. 139.
13 Haug, *Critique of Commodity Aesthetics*, p. 55.
14 Ibid., p. 16.
15 Ibid.
16 Ibid., part III, section 4.
17 J. Baudrillard, *The Consumer Society: Myths and Structures*, London, Sage, 1998 [1970].
18 J. Baudrillard, *For a Critique of the Political Economy of the Sign*, trans. C. Levin, St Louis, Telos Press, 1981 [1972].
19 Veblen's book *The Theory of the Leisure Class* (1899) was only translated into French in 1969.
20 Baudrillard, *For a Critique of the Political Economy of the Sign*, p. 30.
21 Ibid., pp. 30–1.
22 Ibid., p. 144.
23 See Thomas Wex, "Ökonomik der Verschwendung. Batailles 'Allgemeine Ökonomie' und die Wirtschaftswissenschaft," in Andreas Hetzel and Peter Wichens (eds.), *Georges Bataille. Vorreden zur Überschreitung*, Würzburg, Königshausen & Neumann, 1999.
24 See my essay *Das Ende der Üblichkeiten. 30 Jahre danach. Eine ethische Revolution als Umbruch der Sitten*, Freitag 48, 10 November 1998.
25 H. Marcuse, *Eros and Civilization*, London, Abacus, 1969 [1955], p. 114.
26 Ibid., p. 115.
27 Ibid., p. 114.
28 Ibid., p. 49.
29 P. Bourdieu, *Distinction. A Social Critique of the Judgment of Taste*, trans. R. Nice, Cambridge, MA, Harvard University Press, 1984 [1979], pp. 106–9.
30 Horkheimer and Adorno, *Dialectic of Enlightenment*, p. 137.
31 Marcuse, *Eros and Civilization*, p. 49.
32 G. Böhme, *Anthropologie in pragmatischer Hinsicht*, Frankfurt/M., Suhrkamp, 1994.

Part II
Aesthetics of nature and art

8 Aesthetic knowledge of nature

The new aesthetics

Ecology is a science, but the ecological is a symbol, a signal. For some, it is a signal to make a turn-around, for others, it is the harbinger of a better future; in any case, the ecological provides an occasion for reflecting critically upon and for revising received thought and behavioral patterns. Thus, the admission of ecological questions into aesthetics has led to a fundamental revision of its basic notions. In opening up the narrow scope of the latter, a huge and until now, in part at least, unknown and untilled field comes into view.

From an ecological perspective, the relationship between the qualities of environment and disposition become central interests for aesthetics. It is a question of the being (*Dasein*), the presence of things, of works of art, of animals, and human beings; being as felt presence on the part of traditional objects and being as the feeling of presence, as disposition, on the part of the traditional subject. What links these two aspects has been called "atmosphere." Atmosphere, as that which emanates from things and human beings, that which fills space with emotional nuances, is at the same time that which the subject participates in by finding itself in such and such a disposition, by becoming aware of its own presence. By making atmospheres its central topic,[1] aesthetics becomes what it really is: the doctrine of *aisthesis* (sense perception). Perception qua *aisthesis* is more than determining data and situations. Perception is a state of stimulation, an *energeia*, a state of actually being: through perception we become aware of ourselves as being present in an environment. Perception is a shared actuality. It is common to the subject and the object, to the perceiving one and the one perceived. The perceiving subject is actual in participating in the presence of things; the perceived object is actual in the perceiving presence of the subject.

What is thereby overcome has been identified: the self-determined limitation of aesthetics to judgment and rational discourse, the repudiation of sensibility (*Sinnlichkeit*) and emotional participation, the repression of the body, the restriction of interest to art and works of art, the dominance of semiotics, and the preponderance of language. What can be gained is becoming evident: the reconstruction of a complete concept of perception; the rediscovery of body presence; a widening of interests, which include the aesthetic value of our life-world; the introduction of the concept of aesthetic labor, that important part of societal activity that serves not production but staging and presenting, of which artistic creation is only the smallest part.

But ...

As a topic, nature was lost to sight in the process of this plunge into the new. The ecological aesthetics of nature was motivated by the problems we have with nature; more precisely, by the problems we have with ourselves in our relationship to it. Is the new aesthetics, wearied by the turmoil of the aestheticization of our world, to lose sight of its initial question, that of an aesthetic thematization of nature?[2]

Nature

The avant-garde and Futurist contempt for nature is now far behind us. Nature returned as a theme in art a long time ago, but not as a model of mimesis, or as a bearer of symbols, or even as a memory or harbinger of reconciliation. Environment-critical art takes one of two stances: it accuses those who destroy nature, or it demands the protection of nature; Land Art investigates nature by articulating the landscape so as to render it more "expressive," or it creates landmarks which draw attention to the landscape in a particular way. "Elements Art" – art which utilizes any of the four elements of fire, water, air, and earth – conveys a sensitive reappropriation, while Minimal Art and "Material Art" – art which presents materials as such – grope their way through classifications – earth, wood, wax, pollen – to the pure "this here." Since Adorno, the notion of nature has been rehabilitated within art theory, too, and contrary to Hegel's long established verdict, recognized as an aesthetic topic not by virtue of its being an artistic product, but in and of itself. It was Martin Seel[3] who achieved this – using the traditional notions of the beautiful and the sublime, nonetheless. Yet for all that, nature remains strangely absent. Land Art does not differ in principle from art in rubbish dumps (see "Haldenereignis Bottrop"); "Elements Art" is undeniably linked to Kükelhaus' sense of pedagogics; "Natural Minimal Art" does not usually go beyond contingent botanizing, as for example, in the gathering and presenting of pollen. And wherever aesthetic theory conceptualizes sensation in relation to nature, we find the sentiment: "It doesn't have to be nature." Compared to the detailed knowledge of nature evident in Hegel's aesthetics – even though he didn't acknowledge it within his field, as a topic – the degree to which it figures in aesthetics today is very slight.

The decisive questions remain unanswered: why do human beings experience nature as beautiful or sublime, as exuberant, melancholic, serious, pleasant, grotesque, or radiant? Can these experiences be accounted for in terms of the characteristics of nature, of its objects themselves? Is there a basis in natural history for these aesthetic experiences? Why are natural forms always contingent to the arousal of aesthetic need? Is it, indeed, the case, as Kant suspected, that in the final analysis, aesthetic need is a need for nature?[4] And, finally, what aspects of nature do human beings experience aesthetically? Or better still: as what do they experience it? Is there a specified aesthetic knowledge of nature?

Aesthetic knowledge of nature

Alexander Baumgarten justified aesthetics as the theory of a specified type of knowledge – of sensory as opposed to rational knowledge.[5] His project failed for two reasons: first, because aesthetics was quickly reduced to a theory of the fine arts and of works of art, and the question of aesthetic knowledge was thereby suppressed, and reappeared as the question concerning the truth of art. Second, as a project, Baumgarten's aesthetics was unable to say anything concerning things that are specific to aesthetic knowledge. Although he posited aesthetic knowledge as knowledge of the individual, and thus asserted the richness of knowledge in relation to the abstract universal, he did not offer anything that could not have been approximated by a gradual specialization of rational knowledge. If we want to reiterate Baumgarten's project within the framework of the new aesthetics, then we have first to ask whether the aesthetic theory of nature can know something of nature that is fundamentally different from what natural science knows, or whether this theory can recognize nature as something fundamentally different. It is impossible to answer such a question, of course, if one views natural science in terms of the richness of its historical forms, and its entire, inestimable potential for development. However, one can also perceive it as a particular type of knowledge if one follows its prevailing self-stylization. Hence nature or, more precisely, the objects of nature, are the topic of natural science not in their sensory "givenness" for human beings, but in their mutual effect on other objects of nature, especially on those which are made into instruments by human beings. What is considered natural-scientific data is not the sensation itself, but rather what can be displayed by an apparatus. This is also sensorily perceived by human beings, but not precisely in the form of sensations, rather in that of symbols, usually in numbers.

My thesis is that nature as a partner of human sensibility is not a topic for natural science (although some believe it is). Thus, the question again arises as to whether something special about nature is revealed through the agency of human sensibility, or whether nature *itself* is revealed as something special. To affirm the latter part of this question, it may be sufficient to call upon that great tradition in the conception of nature represented by the likes of Aristotle, Goethe, and Alexander von Humboldt. Yet it is a suppressed and, it seems, disdained tradition. The common inability to understand this tradition is due not only to inexperience in transcendental thinking, but has its roots in natural-scientific realism. Here, perception is conceived merely as a more or less satisfactory instrument for grasping nature as it factually is; perception is not seen as a special mode of access to nature, a mode which in itself renders a special appearance to the nature apprehended. Even where, today, such an idea is possible within perspectivism or radical constructivism, the idea that it may be essential for nature to be perceived is not within the range of discussion. But this is precisely the message of this tradition: nature is aistheton, the perceivable. Let me recall the most succinct examples of this conception of nature.

For Aristotle, nature is what is stated by the term *physis*: germinating, blooming; that which reveals itself. That is why, in its essential aspects, nature is characterized by how it corresponds with possible perception. The four elements – fire, water, earth, air; the primary elements of all matter – are determined according to the qualities of warm, cold, moist, and dry. This is completely incomprehensible for us post-Cartesians, for these are secondary qualities; "subjective," as we would say. Yet it is not by a quasi-extraterrestrial, disembodied authority that these qualities are judged, it is by the human as a bodily being, a being which depends on a metabolism; that is, the human being insofar as it is nature itself. Above all, perception is the sense for food. To the extent that the elements are seen in relation to the qualities hot/cold, moist/dry, they are conceived of in terms of the potency attributed to them in the natural context of life. Perception here is an actuality not only for the one who perceives, but also for the thing perceived, the actualization of that potency. In morphology, the metamorphosis of plants and in his theory of color, Goethe not only outlined but also carried out a natural science that remains decisively within the parameters of the phenomenal: "Don't look for anything behind the phenomena, they themselves are the theory."[6] In the conception of nature which was prevalent in modernity, form and color seem to be something incidental in nature, warranting attention only to the extent that they can be justified by having some function. Where a phenomenon doesn't even have a physical presence, as in the case of color in shadows, for example, it seems impossible for it to be granted serious consideration. In the end, the sensory/ ethical effect of colors could only be understood in terms of symbolism and convention, that is, as something that might be left to psychology and the cultural sciences (*Geisteswissenschaften*), but never to natural science.

Goethe, by contrast, pursued a conception of nature in which expression is relevant, and in which color is a phenomenon between subject and object; an actual fact in which the visible and the seeing eye unite. Colors are the acts of light, as Goethe says; *energeia*. Finally, Alexander von Humboldt advanced a plant and landscape physiognomy in his *Ansichten der Natur*, and later in *Kosmos*. Here he not only followed Goethe, but also the theory of garden design as it was presented in Hirschfeld's monumental work. He called for a knowledge of nature which, by grasping certain *physiognomical* features of the character of plants or of the landscape, could render an account of the "total impression of a locality." By "total impression," he meant not so much the quantity of the perceivable in its totality, as the "mood," the "pleasure," or the "magic of nature": the corporeally emotional disposition into which one falls when spending time in the countryside. Humboldt, the travelling natural scientist, paid serious attention to an aspect of the experience of nature that had no place in prevailing natural science, especially in its mode of representation, namely, the bodily presence of the scientist in nature or among the objects of nature. It was in this respect that he wanted to consider the work of painters: here he was following, correctly, his insight that, by virtue of their skills – by being able to create atmospheres by means of concrete attributes, that is to say, with form, color, and composition – painters can convey the sensory experience of nature. Their task, then, is to "grasp and reproduce ... the total impression of a locality."[7]

Aristotle, Goethe, and Alexander von Humboldt; all three sought a conception of nature that would acknowledge the essentially constitutive role of perception itself: it is a basic feature of nature to be perceived. This notion always conflicted with other conceptions; nature as a context of movement, or as a context of the interplay of forces, or nature as a hierarchy of self-organizing entities. It was nonetheless supported by an author whose cryptic influence, mediated by Oetinger, can be demonstrated in Romanticism and throughout the time of Goethe: Jakob Böhme. Böhme conceived of the individual thing as a musical instrument. Its essence is destined to sound, and its material attributes are its "signature," that is, they articulate, like the tuning of an instrument, its "utterance form." Every instrument has its special sound, every flower its special smell. What we try to grasp as the interplay of forces is, according to Böhme, a reciprocal articulation and resonating of things. Nature as a whole is compared to an organ; its "weaving and living" to a huge concert. This idea of nature is not so difficult for us to understand, today. It is about to leave the cryptic theosophical tradition in which it is embedded, and become part of today's prevailing conception of nature:[8] nature as a context of communication. From the investigation of molecular processes to research on animal behavior, today, natural science itself suggests that it is in communication that we find a basic feature of nature. Thus, that which enables the things of nature to enter this context is seen as an essential component of their being nature.

Communication in nature

There have been times when human beings have thought it necessary to distinguish themselves from other living beings, on the basis of their capacity for language – may they be long past. The practical gap between man and animals has become so great that the need for definitive differentiation has disappeared. Progressive anthropology today sees the essential aspect of being human as the capacity for self-styling, by means of which we can disassociate ourselves from our empirical "givenness." Nevertheless, what takes place between nonhuman beings generally remains unrecognized as "communication," it is merely mutual manipulation or functional ritualization, the activities of a selfish gene that has enjoyed evolutionary success. Challenged by the wide range of animal communication, what is accepted as true communication – that is, human communication – is reduced to an ever more complex and ambivalent competence, such as irony or lying.

The number of examples of communication in nature, demonstrated by scientists from fields as far-ranging as molecular genetics and ethnology, is indeed overwhelming.[9] Take warning cries, used between the members of one species or even between different species, which signal the danger of an approaching predator. Some of these signal cries, those of the vervet monkey, for example, convey not only the presence of a predator, as such, but also what kind of predator it is.

Everywhere in the animal kingdom, especially in the context of reproduction, communication occurs concerning the identification of species, groups, family, and even neighborhood membership. Individual recognition between parents and children is very important for the rearing and care of offspring, as is the case with penguins and sheep. Diversely modulated and coded sounds are used to regulate the phases of cooperation, group formation, or couple behavior.[10] The Douglas Bark beetle, for example, emits five different rattle sounds: to deter other beetles from entering the feeding cavity, to signal aggression toward rivals, to conduct courtship, to vent stress, and, finally, to keep other females burrowing and feeding in the same trunk at a distance.[11] On the one hand, there is the idle chatter of gorillas; vacuous but calming, it helps to maintain stability within the group;[12] on the other hand, in the language of bees, precise information is imparted on the direction, distance, and quality of a feeding place.[13] But signals are also utilized deceptively: there is a predatory fish that imitates the gestures of the so-called Cleaner Fish, in order to gain access to a host; there are beetles that smear themselves with the nest odor of certain ants, so that they can be fed by these ants; there are fish species whose smaller males behave like females in order to be able to participate in the insemination carried out by the larger males.[14] Finally, there are the many forms of protective mimicry: flies with the physiognomy of wasps, and moths whose open wings depict the glaring eyes of a glaring owl. Everywhere, there is the communicating of disposition and threatening behavior, not to mention the conveying of an inner state: the baring of fangs, bristling hair, swollen crest, inflated checks, the appearance of colored spots on the forehead and temples.

Communication, signal transference, information processing, coding, and understanding – these take place not only between more developed animals but, through unicellular organisms, all the way down the scale to the molecular processes of the individual cell. By means of a luring substance (*akrasin*), mucus fungi organize into colonies and form differentiated, soporiferous organisms.[15] The messenger RNA absorbs and transports information in the cell. The ribosomes process this information by reading it as an instruction to produce protein.

What does all of this mean? Has the introduction of the concepts of communication and information changed our notion of nature? Or is it just a matter of the metaphorical application of terms which is implemented in a methodology of "as if"? Is it, for instance, just convenient for us human beings to talk about certain processes in nature *as if* they were communication?

To counteract such reservations, we can advance a criterion that dis-associates all the examples mentioned from a concept of nature as a pure context of the interplay of forces:

Telling someone to jump from a bridge is different from pushing him or her.[16]

As an aspect of nature, communication is at odds with the traditional concept of the interplay of forces, in that the energy transfer is relevant to the effect not in its magnitude but rather in its form, its modulation. The difference thus posited in the relationship between cause and effect, that is, between sender and receiver respectively, is even more pronounced in that the law of conservation applies to energy but not to information. Information can be lost, but information can also be generated anew and expanded.

Researching the communication and information processes in nature adds a new characteristic to our image of nature. Though it is first and foremost a matter of processes in the context of life, this suffices to qualify the statement that *communication is a basic feature of nature*. In pursuing this perspective, natural science is about to fundamentally transform itself.

The ecstatics of nature

Newton once established, in contradiction to Descartes, that besides extension and impenetrability, the body's capacity to arouse our senses and our imagination is to be counted among its essential attributes[17] – an amazing observation, in the light of mainstream modern thinking. That bodies can attract and collide seems self-evident, but that they are able to produce images of themselves – at that point we post-Cartesians become hesitant. We are more willing to understand ourselves in terms of things that can be attracted and collided with, and that generate images of an external world because of this interplay of forces.

But if communication is a basic feature of nature, then it has to be conceived of as such; then it is not enough that there are organs of perception, there have to be organs of self-revelation, too, or as Adolf Portmann says, organs of self-presentation.[18] Sender and receiver, demonstration and perception, are all constitutive of communication, just as the speaker is constitutive of the hearer. In generalizing, Buytendijk spoke of the "demonstrative existence value" of things. It is precisely this that aesthetics discovers in nature. Sensory knowledge corresponds to the things of nature insofar as they step out of themselves, present themselves. Aesthetics, as sensory knowledge of nature, recognizes these things of nature in their *ecstatics*.

So what does recognize? In what way does nature step out of itself? What are organs of self-revelation? Modern natural science has long liberated things from their Cartesian limitation, from being restricted to their volume. Substance was dissolved into force, bodies into fields. But that is not enough. The difference between energy and information makes that clear. Furthermore, although in the *sphaera activitatis* one can sense the presence of something, what this something is remains undetermined. Articulation is necessary; modulation of the energy emitted. The pauses in a signal are just as important as the surges of energy – perhaps more important. Nature forms organs of self-revelation by means of articulation, modulation, distinction, signatures, and the formation of models. In this way, the things of nature acquire a physiognomy and an attractive appearance. The aesthetic relationship to nature consists in entering into the physiognomy of things, in being told something by them. Sensory perception means participating in the articulated presence of things.

According to Hegel, the idea is locked within nature. It generates unity in the things of nature by means of organization, but it does not step out of itself. "What appears is only a *real* totality whose innermost invigoration remains behind as an *inner* totality"[19] – hence, in his opinion, the reason for the deficiency of natural beauty. But is it true, then, that "the individual in this sphere (of nature) does not (render) the appearance of independent and total liveliness and freedom, on which the concept of beauty is based"?[20] Was Hegel blind? In any case, we have the task of placing degrees of self-revelation alongside Hegelian degrees of unity in nature. The aesthetics of nature would not merely be the starting point that real aesthetics moves off from, but its true foundation.

The cipher of nature

In *The Critique of Judgment*, Kant inquired into "the true interpretation of the cypher in which nature speaks to us figuratively in its beautiful forms."[21] It is customary not to take Kant's reference to "cypher" very seriously, even though the term was widely used in Romanticism.[22] Likewise, insufficient attention is paid to Kant's remark that the "modifications of light (in coloring) or of sound (in tones) ... embody as it were a language in which nature speaks to us and which has the semblance of a higher meaning."[23] As Goethe refers to the sensory–ethical effect of colors, so Kant attempts to account for the fact that colors in some way "attune" us to certain ideas, that is, they put us in a certain mood. Such efforts are disregarded because reference to the cipher or language of nature does not seem legitimate. Kant himself retracts his remarks by giving them the status of analogy, "as if... The bird's song tells of joyousness and contentment with its existence. At least so we interpret nature – whether such be its purpose or not".[24]

Have we gone beyond this predicament? Are we closer to a true interpretation of the cipher of nature? One thing is clear: by regaining the full concept of perception – in terms of an affecting and bodily "finding" of oneself in an environment – we are also enabled to say in what sense nature reveals itself as a *correlative* to such perception: nature, or more precisely, natural beings, are experienced in their ecstatics. They are experienced as that which steps out of itself, as that which develops organs of self-presentation in the course of evolution. Natural beings are experienced in their articulated presence.

In terms of a theory of nature, perception itself has to be understood as a counterpart to this basic feature of nature: to be ecstatic. The idea that natural beings step out of themselves is older and more advanced than that of perception. It may indeed frequently be the case that the articulation of pre- sence and perception are coeval, as in the evolution of flowering plants and insects. However, as Portmann correctly maintains, there is also non-addressed self-presentation: "They are patterns that are simply there, without any use, without any purpose."[25] He points to the haloes of sea anemones, the patterns on the skins of snakes, the patterns and outgrowths of sea snails. An aesthetic principle which posits that a particular type of perception corresponds to each form in nature (M. Hauskeller) can only have regulatory significance. Human perception seems to go beyond being merely a functional correspondence of perceiving and acting, as believed by biologists like Jakob von Uexküll, to the extent that it is also receptive to the non-addressed self-presentation of natural beings. This means that human beings can be emotionally touched by (as Kant says, attuned to) the forms and charms of nature, even where they are of no adaptive value to them as Irving beings. This insight may be behind Kant's reference to a disinterested delight in beauty. But in being emotionally touched by nature it is a matter not simply of beauty and the sublime, nor of a moral feeling, but also of completely different atmospheres.

By understanding nature as ecstatic, and through its rediscovery as *aistheton*, we can justify the use of concepts such as the "cipher" and "language" of nature. What these terms refer to is the articulation of natural beings in their presence. But whether it is a matter of script or language, whether metaphors of *litterae* are appropriate, seems questionable here. For what Kant calls the "interpretation of the cypher" of nature proceeds in an opposing direction to that required by metaphors of script and language. It does not follow the path in the usual direction: from a clear knowledge of letters to the discovery of their meaning, but – if expressed in these terms – inversely, from the meaning experienced to the discovery of the letters. That is why it is more appropriate to speak, as Klages does, of the actuality of images, or – like Schmitz – of atmospheres. What is experienced when faced by the things of nature or within environments, is primarily the atmosphere which they emit, which takes hold of us, the perceiving ones, which puts us in a certain mood, which enwraps us. It is only from the vantage point of this experience that it is possible to ask about the patterns, the "forms and charms," the articulations, the physiognomy of natural beings, in order to render an account of these experiences. Which patterns, lines, contrasts, forms, and colors are relevant can only be discovered in this way; they are a specific subject matter of an aesthetic theory of nature.

Aesthetics recognizes nature in its ecstatic. Through perception we enter a common actuality with the things of nature. Thus, just like other atmospheres, the beauty of nature is also an actuality of human beings.

Notes

1 See Gernot Böhme, "Atmosphere as a Basic Concept of a New Aesthetics," *Thesis Eleven*, 36, 1993, 113–26.

2 On this outline of an ecological aesthetics of nature, see my publications: *Für eine ökologische Naturästhetik*, 2nd edn, Frankfurt/M., Suhrkamp, 1993, and *Natürlich Natur. Über Natur im Zeitalter ihrer technischen Reproduzierbarkait*, 2nd edn, Frankfurt/M., Suhrkamp, 1992.

3 M. Seel, *Eine Ästhetik der Natur*, Frankfurt/M., Suhrkamp, 1991. For a critique, see H. Böhme's book review in *Zeitschrift für philosophische Forschung*, 1992, pp. 319–26.

4 "art can only be termed beautiful, where we are conscious of its being art, while yet it has the appearance of nature." Immanuel Kant, *Critique of Judgment*, trans. J. C. Meredith, Oxford, Clarendon Press, 1952, p. 167, § 45.

5 H.-R. Schweizer, *Ästhetik als Philosophie der sinnlichen Erkenntnis*, Basel, Schwabe, 1973, esp. p. 46ff.: "The importance of rhetorical concepts for the truth of individual appearance."

6 J. W. von Goethe, *Scientific Studies*, New York, Suhrkamp Publishers, 1988, p. 107.

7 Alexander von Humboldt, *Kosmos*, vol. 2, Stuttgart, *Gesammelte Werke*, 1844, p. 66.

8 On this see my article on Jacob Böhme in Gernot Böhme (ed.), *Klassiker der Naturphilosophie*, Munich, Beck, 1989, pp. 158–70.

9 R. Dawkins and J. R. Krebs, "Animal signals: Information of manipulation?," in J. R. Krebs and R. B. Davies, *Behavioural Ecology. An Evolutionary Approach*, Oxford, Blackwell, 1978, pp. 282–309.

10 T. A. Seboek (ed.), *How Animals Communicate*, Bloomington, IN, Indiana University Press, 1977.

11 L. C. Ryker, "Kommunikation beim Douglasienkäfer," *Spektrum der Wissenschaft Biologie des Sozialverhaltens*. Heidelberg, Spektrum, 1988, pp. 116–24.

12 Hess, *Familie 5, Berggorillas in Virunga-Wäldern*, Basel, Birkhäuser, 1989.

13 M. Lindauer, *Verständigung im Bienenstaat*, Stuttgart, Fischer, 1979.

14 T. Halliday in particular deals with deceptive communication in the animal kingdom, in his article "Information and Communication," in T. R. Halliday and P. Y. B. Slater (eds.), *Communication*, Oxford, Blackwell, 1983, pp. 43–81. See also J. E. Lloyd, "Die gefäschten Signale der Gühwürmchen," in D. Franck (ed.), *Spektrum der Wissenschaft "Biologie des Sozialverhaltens,"* Heidelberg, 1988, pp. 96–104.

15 J. T. Bonner, Lockstoffe sozialer Amöben, Spektrum der Wissenschaft, "Biologie des Sozialverhaltens," Heidelberg, 1988, pp. 74–80.

16 Dawkins and Krebs, "Animal signals: Information or manipulation?," pp. 282–309. The example is said to come from Cullen.

17 I. Newton, *Über die Gravitation*, annotated and trans. by G. Böhme, Frankfurt/M., Klostermann, 1988, pp. 56–57.

18 A. Portmann, *Farben des Lebendigen, Palette* (anniversary issue), Sandoz (1886–1961), pp. 4–22. They are called "organs of announcement" in his book *Das Tier als soziales Wesen*, Zürich, Rhein, 1953.

19 G. W. H. Hegel, *Werke in zwanzig Bänden*, vol. 13, *Vorlesungen über die Ästhetik I*, Frankfurt/M., Suhrkamp, 1970, p. 195.

20 Ibid., p. 198.

21 I. Kant, *The Critique of Judgment*, § 42, p. 160. See my extended interpretation of this paragraph in Landeshauptstadt Stuttgart (ed.), *Zum Naturbegriff der Gegenwart*, vol. II, Stuttgart, Frommann-Holsboog 1994, pp. 39–58 (translator's note: Meredith translated *Auslegung* in this passage as construction, whereas we favor interpretation in this quote).

22 A. von Bormann, *Natura loquitur. Naturpoesie und emblematische Formel bei Joseph von Eichendorff*, Tübingen, Niemeyer, 1968.

23 Kant, *The Critique of Judgment*, § 142, p. 61.

24 *Ibid.*, § 42, pp. 161–2.

25 A. Portmann, *Farbige Muster im Tierreich*, CibaRundschau, vol. 14, 1963, p. 10.

9 Nature in the age of its technical reproducibility

Introduction

In 1936 Walter Benjamin published an essay which, particularly after its appearance in German in 1955, was to have a decisive influence in the debate over the difference between classical and modern art. It was titled *The Work of Art in the Age of Its Technical Reproducibility*.[1] In this essay, Benjamin notes that art – he is thinking primarily of visual art – is affected fundamentally, at its "core," by the fact that it can be technically reproduced. In talking of the "core," he means that it is not merely the manner of producing art that is affected – as in the case of modern painting, whose development cannot be considered without taking into account the impact of photography as a competitor. He means that the very essence of the artwork is changed, and he even asserts that this change affects all existing works, including the *Mona Lisa*. Benjamin does not expand on the techniques of reproduction that he has in mind, though he must at that time have been thinking of photography, film, and the gramophone record. However, in referring to *technical* reproduction, he clearly means something other than traditional copying by craftsmen. He is envisaging a perfected technique which does not merely produce a further example of a type, but actually replicates the individual work. What such techniques endanger, according to Benjamin, is the "genuineness," the uniqueness, and unrepeatability of the original, and thus also its dignity.[2] By 1936 Benjamin may have had a premonition of the scale on which the original works of European art and history were soon to be destroyed. We, at any rate, have grown accustomed, since 1945, to taking reproductions for originals in many places – for example Hildesheim Cathedral, or the market square in Warsaw. For Benjamin, the mere possibility of their reproduction caused artworks to lose their special aura,[3] the halo that gave the classical work its historic individuality and conferred upon it the status which commanded respect.

This change was taking place – as Benjamin well knew – not within the individual artwork, since the aura of a work is not, of course, something that is physically attached to it. Rather, the loss of the aura was one aspect of a change in art as a social institution, and in the social function of works of art – a change that went hand in hand with the industrialization of art production. By the mass replication of artworks, or by even the mere possibility of replicating them, the manner in which they are perceived is fundamentally altered. Benjamin speaks of a substitution of exhibition value for cult value in the work of art.[4]

This transformation of art as an institution, that Benjamin explained with the technical reproducibility of artworks, was deliberately promoted by the avant-garde artists of the time. They sought to abolish art, or to redirect it into life. The means they used were sometimes the same – as in Duchamp's ready-made objects – and sometimes different, but in all cases they aimed both to shatter the aura and to make visible the work and the production process that gave rise to it. What the avant-garde frequently overlooked, however, and what Benjamin explicitly maintains, is that these techniques and strategies would not lead to the abolition of art; they would merely put an end to the traditional concept of art and give birth to a new one. Benjamin shows in the essay I have referred to that the reproduction techniques which mean the death of the original and thus of the artwork in the classical sense, become the *production* techniques of artworks of a new kind. He is thinking of productions that do not involve any original, since the product appears from the first as a kind of apparatus, something intrinsically dependent upon techniques of reproduction. Such is film and such, too, is a large part of modern music: the part which is technically synthesized. Benjamin cherished the hope that this synthetic quality might liberate art from the esoteric realm of cultivated middle-class enjoyment and professional criticism, and make it a component of mass culture.[5] He realized that it would then be drawn into political life, whether through the aestheticized politics of Fascism, or through the politicized art of Communism.[6]

Reading Benjamin's essay *The Work of Art in the Age of Its Technical Reproducibility* today, we are bound to feel a mixture of agreement and dissent, since the basic social choices present themselves to us differently now. But we also feel, surely, the sense of shock which Heidegger called the "basic mood of philosophical thought in the twentieth century."[7] We have an inkling that in his essay, Benjamin may have expressed a truth about our time that goes far beyond the niceties of art-historical debate. This truth becomes apparent if we experimentally substitute the word "nature" for "artwork" in the title of Benjamin's essay, or if we merely realize that – as art and nature have been twin concepts in our culture since the ancient Greeks – neither can undergo a fundamental shift in its meaning without the other.

The decay of aura: in art this meant the abolition of aloofness and awe, the tendency to destroy uniqueness, the loosening of traditional bonds and the shift toward functional use and exchange values. In nature, too, it can mean the refusal to acknowledge a given world, the loss of reverence to life, the annihilation of individuality, and it certainly implies the radical abuse of nature as a commodity. What shocks us in the face of such possibilities, jolting us out of benign, Romantic attitudes toward nature, is the renewed awareness that in talking about nature we are talking about ourselves: the nature that we ourselves are. The technical reproducibility of nature calls into question our basic assumptions about ourselves.

Hubris?

"The technical reproducibility of nature": we shrink from even using the words. Man as the second creator of nature? Is it not presumptuous to suppose that nature could ever be so controlled that we would be responsible for its continuing existence? Is nature not the encompassing medium that we must presuppose in everything we undertake?

Nature is always greater than the human being. Every enhancement or destruction of nature that we bring about remains within its framework. But does this imply, conversely, that everything that human beings do to nature, or turn it into, is irrelevant? It seems that we need to find a proper perspective from which to judge.

If we are obliged, today, to talk of nature as a social and historical product of human beings, we mean the nature that is here on earth: what used to be called the sublunary realm; and we are applying a scale commensurate with the human being. It is the nature which is relevant to us, but it has not yet been properly conceptualized.[8] The manipulation of nature by mankind is never unlimited or unconditional. Human beings are always dependent in some way upon a given nature, upon a particular "material," and they have to make use of existing natural laws. However, the part of nature which has to be taken as given is noticeably disappearing, with the escalating advances of technology. From materials that are formed by hand from wood and metal, nature's presence has receded to the elements that are used to form plastics, amalgams, crystals, and then to the elementary particles and atomic fragments that are forced together and reconstituted by fission and fusion, and hybridized as isotopes.

What of the laws of nature? Of course, man is still dependent upon the conservation of energy, gravity, and particular reciprocal effects, but beyond that, the major element in regular, deterministic behavior in synthesized nature is its organization.

We must ask, therefore, how deeply the human reproduction of nature penetrates our world, and how far its theoretical significance extends. To get an idea of the dimensions involved, let us indulge in a little science fiction – just a little, to explore the tendencies we see. We should think first of the oldest realms in which the reproduction of nature was not left to itself, but was partly shaped by human labor: agriculture and forestry. The fundamental turning point here came with Liebig and the beginning of industrialized agriculture. Since the advent of Liebig's artificial fertilizers, agriculture has no longer been a natural cycle of reproduction, loosely guided by human activity, but the reproduction of a particular natural function, of a more or less specific fertility, through the introduction of substances derived from outside the regional cycles. The tendency in this development veers toward "soil-less agriculture"; the production of large-scale generators which convert a regulated supply of nutrients into the desired biomass, without involving any spontaneous vegetation.

Second, let us consider the case of Irving organisms, and in particular, of species. Species were regarded by the ancient Greeks, as they were in the Christian worldview, as a framework that had existed for all time, or that had been instituted by God: nature as a datum. To be sure, as long as it has existed, mankind has bred races and variations upon them, including those which could not have continued to reproduce without human intervention. Ever since Darwin advanced his theory of evolution, we have been aware that species, too, are in a state of flux, and yet even by the terms of this theory, existing species have to be accepted as given by nature. Now, however, human beings are beginning to intervene productively in the evolutionary process; not merely by selecting variants and thus guiding natural reproduction, but by deliberately producing and reproducing variants: the end product is the new living organism, designed to fulfill specific functions, and patented for the benefit of its producer.

Let us consider, third, the so-called ecosystems. These are regional networks of organic and inorganic processes that reproduce themselves cyclically – and are only modified over long periods, by evolutionary developments. We have long been aware that ecosystems can no longer be regarded as anything other than ideal constructs, and that the natural state of these processes no longer reproduces itself anywhere either regionally or globally, in a manner that is not influenced by human actions. They can therefore be referred to more realistically as "humankind-organized ecosystems"[9] or, still more modestly, as ecological constructions.[10] The preservation of a desirable natural environment or of regional partial environments requires of human beings an ever-growing input of work, external energy, substitution of materials, and regulation of the entire network. If this line of development were extended, it would lead to a perfected eco-management whereby the stabilization of natural systems would no longer be self-regulating but computer-controlled. The excessive demands imposed by such a development have already given rise to attempts to repro-duce nature on a different basis, that is, to restore natural mechanisms of self-regulation. The spectrum of these forms of the technical reproduction of nature ranges from the re-naturalization of streams and the artificial creation of marshlands – complete with all the associated plants, microorganisms, insects, and smaller and larger animals – to the "re-cultivation" of entire landscapes devastated by industrial use.

Fourth, let us think about the purely aesthetic reproduction of nature. Here, one might say that the appearance of nature is reproduced outside the context of nature. Examples of this range from the tree maintained by hydroponics in the pedestrian precinct, and the underground garden restaurant in subways, to the artificial Christmas tree. The aim of such artifacts is to satisfy the need for natural scenery by gestures evoking an imaginary nature.

Finally, let us consider ourselves, insofar as we too are of nature – our own bodies. Our reproduction, which was once carried out almost incidentally within the course of nature as the accompaniment of love and the convivial organization of mealtimes, is tending to split off from these cultural forms of behavior and to take on a technical regime of its own. When applied to nutrition, this could mean that meals come to be no more than the ritual consumption of well-formed culinary elements that are supplemented before or afterwards by vitamin concentrates, in tablet form. When applied to the production of offspring, it might mean that physical love is, in principle, enacted in sterile bodies or infertile periods, while conception qua the fertilization of eggs is done externally and systematically. Antenatal diagnostics, followed by selection, manipulation and therapy will ensure that the children thus produced are constituted in a desired and "acceptable" way.[11]

Let us also give some thought to the sphere of medicine. Nature, it used to be said, is the best doctor, and the doctor can do nothing if nature does not help itself. However, reproduction in the sense of a return to health has, with the advance of modern medical science and technology, become something that should be brought about intentionally and causally. Examples of the technical reproduction of the human body range from the replacement of regenerative mechanisms by technically induced processes, and organ transplants to artificial limbs. Human beings now technically reproduce their own nature in the form of the computer-controlled prosthesis.

Nature and technology

I do not wish here to trace the consequences of the technical reproducibility of nature at a concrete level. I do not wish to ask how far the developments that are emerging in principle can or should be realized in practice. The dangers inherent in such developments, the ethical questions they raise and the alternatives that might be adopted, are now being discussed from many points of view. Rather, I should like to make it clear that regardless of how far we intend to exploit the possibilities of reproducing nature by technological means, something fundamental has already happened. I wish to show that the thing which is, for us, nature, has already changed in a crucial way, and that the situation of human beings, who are themselves nature within nature, has also changed crucially. The same idea might be expressed by saying that even before its technical reproduction has entirely changed nature for us in a concrete way, nature as an essential element of our European culture has already been devalued, if not to say destroyed. This may be concealed by the fact that "nature" as a dominant cultural idea is enjoying a boom at precisely this moment: nature as a leisure value, nature as an ingredient of consumer aesthetics, nature as a political objective, nature as a seal of quality for everything imaginable. Anyone who attaches the epithet "natural" to his wares intends to signify thereby that they are especially good: natural soap, natural cosmetics, natural colors. A magnified version of the same thing is the epithet "bio": bio-wines, bio-detergents, bio-energy. Nature as a value is a late descendant of a strand within our European tradition in which nature was a normative idea supported by ontological, cosmological, and theological arguments, standing opposed to the ideas of culture and civilization. Expressions like "natural law," "naturalness," "the natural state," "in the nature of things" bear witness to this. In all these expressions an original, given order is ascribed to the world, to each separate object and to human relationships, in contrast to which all man-made arrangements are seen as precarious. Since the Greek enlightenment, the outline of the term "nature" has been defined through contrasts: nature and technology, nature and culture, nature and civilization, nature and law are examples of this.

In such antitheses nature is understood as something which stands opposed to the human realm, in the sense that nature refers to something that exists of its own accord, is founded on itself and as such is reproduced from itself, whereas the human sphere is determined by law and statute, by production, and by the fact that whatever exists and has validity here and now is only maintained through its constant affirmation and reproduction by human beings. A prominent example of such contrasts is the antimony *φυσις/τεχνη*. *Physis* is the Greek term for nature, and refers to that which arises spontaneously, which reveals itself. *Techne* means human expertise, productive knowledge, and refers both to what we now call technology and to crafts and agriculture, but it also includes the realm of art. According to Aristotle, *physis* is the realm of that which is there of its own accord. Something which is or exists by nature is something which has the principle of its motion within itself.[12] By contrast, something which is or exists technically, derives the principle of its motion from human beings and human purposes, and depends on human beings for its very existence and its ability to be reproduced. The famous example Aristotle uses to illustrate the distinction is the willow-wood bedstead. If such a bedstead is buried, a willow tree grows from it, not a bedstead.

Nature, according to this basic idea which has had a profound influence on European culture, is that which exists of itself and reproduces itself. The ancient Greeks thought it existed eternally; in the Christian context it was believed to have existed since the origin of divine Creation. Now it can, of course, be said that this idea of nature has been progressively eroded since the early modern period. As early as the Renaissance, nature as a whole was regarded as a clockwork mechanism. Then, Descartes asked us to understand nature as a craftsman understands his craft, in terms of making and producing. Animals, and finally the human body, are conceived on the model of the machine. That may be correct, but we should not allow it to obscure the fundamental turning point represented by the possibility, in this century, of the technical reproduction of nature.

Even though nature was conceived as the product of a craftsman it was nonetheless the work of a divine craftsman, who "eminently" surpassed the capacities of human crafts, meaning that the two kinds of production were separated by an unbridgeable gulf. Newton relegated what he took to be the non-scientific questions of the order of the planetary system and, still more fundamentally, of the basic properties of matter, to the sphere of theology. Even Kant, for whom Newton's work was a kind of prototype of scientific explainability, regarded a Newtonian explanation of, for instance, a blade of grass, as inconceivable. Liebig, an opponent of the Romantic philosophy of nature and a staunch supporter of experimental science in the nineteenth century, still believed a special life force to be necessary in explaining organic order in nature. I do not need to prolong this list of thinkers who bore witness to the classical concept of nature. It is clear that we have fundamentally departed from such ideas, in view of the synthesis of DNA sequences, the technical production of elementary particles, and the construction of new elements. It is not simply that in principle we see no limits to our ability to produce whatever is found in nature; it is not just that we now meet the Cartesian injunction to perceive nature in terms of the way it is produced not merely figuratively but literally – but we have reached a point where nature is no longer something simply given: nature is anything which, in principle, it is possible to produce.

"Nature is that which it is, in principle, possible to produce." This proposition is clearly paradoxical in relation to the classical concept of nature, and, indeed, it takes that concept *ad absurdum*. Equally, however, it is absurd not to regard elements with atomic numbers in the region of 100 as natural, or to refer to human insulin produced not in human beings but through genetically modified bacteria as anything other than natural. And it is no longer possible to draw a clear dividing line between the synthetic materials produced by polymerization processes invented by man, and those generated by plants.

The possibility of reproducing nature technically means the end of a concept of nature that found vivid definition precisely in its antithesis to the sphere of human production. Present-day appeals to nature as a value are no more than ideology, in that they base themselves on a solid idea of nature at the very moment when – historically, and no doubt irreversibly – that idea is disintegrating.

Nature and art

The traditional antithesis between *physis* and *techne*, between nature and technology or art, implied not only a distinction between these concepts but also an inner relation between them. This is especially clear in the sphere of classical aesthetics. Here, nature and art illuminated each other. As an example of such thinking, I shall quote a passage from Kant's *Critique of Judgment*: "Nature was beautiful when it also had the appearance of art; and art can only be called beautiful when we are aware that it is art yet has the appearance of nature" (§45). This requires some explanation. To begin with, we should not stumble over the use of the past tense in the first part of the quotation: Kant is using it merely to refer back to earlier discussions. He is therefore asserting that we describe nature as beautiful when it looks like art. This "looking like art" means that nature spontaneously displays a degree of order and regularity that we, as Kant says, can only understand as something produced according to a plan, i.e. a product of technique. Kant speaks accordingly of a "technics of nature," as in the formulation: "nature's autonomous beauty reveals to us a technics of nature" (*Critique of Judgment*, 77). This means, however, that we can only comprehend a basic feature of nature that we first perceive through aesthetic experience if we view nature as analogous to technology. In the second half of my quotation from the *Critique of Judgment*, Kant asserts the converse of the above, namely that art can only be thought of with the aid of the concept of nature: "And art can only be called beautiful when we are aware that it is yet has the appearance of nature." If Kant speaks here of "beautiful art," it is to demarcate the field that we now commonly call art – a demarcation that was needed at the time because the term was still used indiscriminately to include all other acts of production such as crafts and technology. It is Kant's contention that in works of beautiful art we cannot see, or are supposed not to see, that they have been made. They should look as if they existed of their own accord, that is, as if they were a part of nature. To confer on something produced intentionally an appearance of autonomy, and on something entirely subject to rules, an air of spontaneity – that, for Kant and the classical view of art, was the essence of art. This idea was closely linked the concept of genius: the artist was seen as a human being in whom and through whom nature produces itself. Classical aesthetics therefore teaches us that nature and technology, although opposed concepts, with features making them mutually exclusive, are on the other hand interdependent, and in important cases, to be conceived in terms of each other. Nature arouses our admiration precisely where it appears to be art, and art only really comes itself when it has left all technique behind and appears as nature. One might add that in the classical view art was seen as an imitation of nature, and that even in theories on technology advanced in our own century, technology has been understood as an imitation of nature.

This intertwining of the concepts of art and nature makes us realize that the concept of nature was necessarily affected by the disintegration of the classical concept of art. On the other hand, the linking between them might also seem to suggest that the technical reproduction of nature does not mark a sudden turning point, as it appeared to indicate in the past, since nature has always been regarded as especially admirable when it has the appearance of something technically produced. This objection overlooks, however, the fact that we find ourselves here *in aestheticis*, that is, in the realm of appearance. Here, the destruction of appearance is the destruction of the thing itself. The important point was that art was not nature, and that nature did not really proceed in a technical manner. Automatic art, in which the artist either tried to allow the unconscious to express itself directly, or produced graphic works by random processes, was one of the factors that brought down the classical concept of art. Another development contributing to its destruction was the fact that modern works of art retained traces of their production, so that the appearance of autonomy, of something existing of itself, was abolished. The aura of art was banished by the disillusioning strategy of making the technical aspect of art visible, and by the explicit use of chance. Likewise, the artist lost his halo, as is made clear in a parable of Baudelaire's, in which a poet, caught up in the bustle of street traffic, has his nimbus cast in the mire. Or, as the Austrian novelist Musil was later to put it, just as the poet had become, more prosaically, a writer (*Schriftsteller*), so the artist had become a mere manipulator of paint (*Malsteller*).

But nature, too, has an aura to lose, a nimbus that once affirmed its significance as a leading cultural idea. This nimbus arose directly from the fact that nature, displaying its own intrinsic order and regularity, looked like the product of masterful technique, which is precisely what, qua nature, it was not. Kant says that this appearance, which constitutes the beauty of nature, has a moral interest for human beings. This moral interest in nature manifested itself especially in the equation of nature with the values "pristine," "good," and "innocent," which became commonplace in the eighteenth century and is still familiar in our day. Nature is honored for its beautiful order, an order that it seems to show effortlessly and spontaneously, whereas civilized man has to strain toward it. The aura surrounding nature or, better, natural objects, is the admirable quality of the order it organizes within itself, its purposefulness without a purpose, as Kant put it. Man reacts to it with a moral attitude, that of respect for a given order.

This aura of nature decays, I have said, if the works of nature are not only thought of as analogous to works of technology, but are actually turned into works of technology. It is this decay of the aura of nature and therefore of nature as a cultural value that we are now experiencing with the advance of its technical reproduction. If I refer to this process as one of rapid disillusionment with regard to nature, I do not mean, of course, that modern science has diminished our admiration for the works of nature. On the contrary, the "techniques of nature" that we have tracked down step by step – one need only think of the sophisticated intercellular protein factories – are far more worthy of admiration than anything the eighteenth century credited to nature. It is only the practical feasibility of reproducing nature that destroys its aura. The experimental approach, that is already bound up with manipulation, itself implies a loss of reverence toward life – as is seen most clearly in experiments on animals. The ensuing practical ability to reproduce natural processes technically, abolishes the respectful distance that previously existed between man and nature, and only the rather overemphatic vigor with which the work is pursued gives an idea of the barriers that have had to be overcome.

As I am concerned at this point in my lecture with the realm of aesthetics, I shall choose an aesthetic example to illustrate the decay of the aura. It comes from the young science of "fractals." Some time ago I saw how the endless superimposition of three rectangles with a fixed standard of reduction gave rise eventually to the general impression of a fern leaf. This fact, that a simple basic shape in conjunction with a recursive generative rule finally produces a very sophisticated and highly "individual" natural form, will naturally give pleasure to anyone with a training in science. And yet one cannot deny the shock one also feels at this manner of producing the "natural" form of the fern leaf.

I should like, too, to give a Kantian example of a similar loss of aura:

> What do poets prize more highly than the enchanting song of the nightingale in a lonely bower on a still summer night by the soft light of the moon? Yet there have been examples when, no such singer being to hand, some jovial landlord has most pleasurably deceived the guests, who had come to his inn to enjoy the country air, by hiding some high-spirited lad who knew how to imitate the natural song of nature ... in a bush. But no sooner were the guests aware of the deception than none of them could endure hearing the song that had just seemed so charming ... it must be nature or be thought by us to be nature, if we are to take a direct interest in the beautiful as such.[13]

This somewhat light-hearted example cannot, of course, really make plain the seriousness of the loss of aura, since it still recognizes a distinct difference between "being nature" and "being thought by us to be nature." What is happening to us is that this very difference is becoming fluid. Through the technical reproducibility of nature, the latter is losing its aura, and is thus being rendered unusable as a cultural value. In our European culture, nature has been a main guideline in orienting ourselves within the cosmos, a foundation stone of our moral awareness and a fixed pole of our legal, social, and political thinking.

I cannot trace here the consequences of the disintegration of the classical concept of nature for all these areas of our culture, and in what follows I shall therefore pick out one important area: that of our awareness of ourselves as human beings. The nature that we are and that we have, has up to now provided a firm starting point for creating this awareness.

The nature that we are

For the purposes of philosophical anthropology, Kant distinguishes between physical anthropology and anthropology seen from a pragmatic perspective. Anthropology from a pragmatic perspective is concerned with the question of what man can or should make out of himself. Physical anthropology is set apart from this, being concerned with what constitutes the physical make-up of the human being, i.e. what is given by nature. Human nature in this sense, our *physis*, is the nature we ourselves are, the body. Everything connected with it and arising from it has hitherto been accepted as a fact, and has been assimilated to the human project of the self in terms of our attitude to it, for example, or through its instrumentalization. One was born as a man or a woman, with a particular disposition and constitution, perhaps even with certain illnesses. The body could afflict us with illnesses, and from it arose the instinctual impulses and moods with which we had to contend. Something similar applied to that which grew out of human physical existence. Children somehow "arrived" as they were; they were a blessing, or they could also be an affliction. The body, one's constitution, children – these were in the first place a fate one had to deal with, but it was in struggling with them that the self found its origin and its strength. Moods and impulses were a challenge to one's self-affirmation, and in coming to terms with them, one acquired self-control and character.

What will become of human beings now that the very thing which made up their nature has been absorbed, in principle, into the sphere of the technically reproducible? How, in future, shall human dignity and respect for ourselves be generated, if the difference between given nature and the self-determining human character is obliterated; if the distinction between facticity and project[14] is blurred?

How is my sense of inescapable concern to be preserved, forcing me to acknowledge that I am my body even if I have it,[15] if, in principle, every organ, even the heart, is exchangeable? How can an unambiguous identification with my sex serve as the starting point of my understanding of myself, if gender can, in principle, be hormonally and surgically manipulated? How are a self and a self-determined character to form, if the struggle with constitution, impulses, and moods is no longer the province of the will but of expert, rational, and functional control by drugs? How is the burden which children always represent to be borne by parents, if they are no longer accepted in principle as something given, or fated; their existence becoming the responsibility of their parents through the use or non-use of contraceptives? How shall they be able to come to terms with their constitutions, their illnesses, their sex, if the possibility of antenatal diagnostics, selection, and therapy bring even the quality of children within the responsibility of the parents?

It has been said that the possibilities of the technical manipulation and the physical reproduction of human beings have made human nature contingent.[16] This might be expressed more trenchantly by saying that anything resembling human nature is thereby abolished, since nature meant precisely the given and fated fact of existence. Of course, it cannot be said that there is no longer anything given, fortuitous, or fated in the self. Each human being will always have to contend with the tension between facticity and project. The decisive change is that one now has control of the borderline between them, so that the human project will finally consist of no more than what one has, or wishes, to accept as a given fact. We can see here, perhaps most clearly of all, the alteration of principle that has been brought into the world by the technical reproducibility of nature and the irreversibility of this event. The disintegration of nature, which here means that which the human being has to accept as his or her physical being, does not depend upon whether or not one manipulates one's body or one's physical reproduction. To make the point with an example: even someone who is not permanently primed with stimulants and sedatives, headache tablets, and psychopharmacological drugs is still required to maintain a constant level of performance. Or, to take a more drastic example, even parents who have consciously decided against prenatal diagnosis can no longer regard a child with Down's syndrome as simply a stroke of fate. That is to say, now that the means are available, everyone will have to decide where to draw the boundary line of nature. Even what is "left" as nature, i.e. is not manipulated, is no longer simply given. Yet that is what man's physical being used to be: the given, the set task.

Conclusion

If we use Walter Benjamin's essay as an analogy for reflecting upon the consequences of technical reproduction, we have to say that in the case of nature as of art, the possibility of such reproduction destroys nature as it once was. By this we mean that nature, prior to all its concrete manifestations, is destroyed as an element of our European culture, is devalued as a guiding cultural idea. However, just as the end of art announced by the Dadaists and the Surrealists, and by Benjamin with them, did not actually mean the end of art itself, no more is the concrete destruction of nature a destruction of nature itself or the disintegration of the classical concept of nature the end of our idea of nature. In conjunction with the development of technical processes of reproduction, a new and different form of art has come into being. What, we might ask analogously, will nature be in the age of its technical reproducibility?

If we attempt to pursue the analogy with Benjamin's analysis of the development of art as it applies to nature, we shall have to proceed very cautiously, bearing in mind that the hopes Benjamin attached to the new art have proved thoroughly deceptive. The new art neither brought a proletarian mass culture, nor did the politicization of art have any results that one can view with pleasure.

What we can say with utmost caution is the following: with the possibility of the technical reproduction of nature, what is understood as nature ceases to be defined by its opposites, by technology, culture, civilization, and the human sphere. This means that nature, as far as it is of practical relevance, must itself be understood as a cultural product, as "socially constituted nature";[17] but on the other hand it also means that humankind, with its culture and its technical possibilities, sees itself more and more as a part of nature. The concept of nature thereby loses its sharp outline and becomes once again, in a sense, a concept of totality, as it was at the time of the pre-Socratic philosophers. This opens up the possibility of a new natural philosophy as a first philosophy, but at the same time closes the possibility of deriving moral norms from the concept of nature. Normative distinctions must be drawn below and within that concept. This undoubtedly means a further, radical disenchantment of the human being's cultural self-image. Whether this new, emerging, comprehensive concept of nature can also give rise to new hopes is difficult to say.

That the hopes Benjamin pinned on the new concept of art proved illusory resulted, we can now say, from the fact that the pair of alternatives that guided his thinking, Fascism and Communism, was a dichotomy that actually obscured the fundamental historical development that was taking place; that is to say, modernization. Our view, too, has been deflected hitherto from the fundamental development in our world by an obscuring dichotomy, the East–West conflict. It is only now becoming apparent that the social and political upheavals of the future will be determined by another set of factors that have long been maturing in obscurity. These are the anthropogenically determined changes to the natural basis of humankind, that is, to the climate, habitability, and fertility of our earth, the effects of which will be mediated politically by the North–South conflict. The politicization of nature has long since begun.

Notes

1 Walter Benjamin, "The Work of Art in the Age of Mechanical Reproduction," in *Illuminations* (trans. H. Zohn), London, 1970, pp. 219–53.
2 Ibid., p. 222f.
3 Ibid., pp. 224–5. On this concept, see Marleen Stoessel, "Aura," *Das vergessene Menschliche. Zu Sprache und Erfahrung bei Walter Benjamin*, Munich, Hanser, 1983.
4 Benjamin, "The Work of Art in the Age of Mechanical Reproduction," pp. 226–7.
5 Ibid., p. 236.
6 Ibid., Afterword, p. 244f.
7 Martin Heidegger, *Beiträge zur Philosophie* (Vom Ereignis), Gesammelte Werke, vol. 65, Frankfurt/M., Klostermann, 1989, p. 21f.
8 This thesis is elaborated in G. Böhme and E. Schramm (eds.), *Soziale Naturwissenschaft*, Frankfurt/M., Fischer, 1985.
9 For example, Prof. H. Sukopp, Berlin, in numerous writings.
10 For example, Böhme and Schramm, *Soziale Naturwissenschaft*.
11 On anthropological changes brought about by technology, see my *Anthropologie in pragmatischer Hinsicht*, Neuauflage, Aisthetis-Verlag, 2010 and *Invasive Technisierung. Technikphilosophie und Technikkritik*. Kusterdingen, Die Graue Edition, 2008.
12 Aristotle, *Physics* B1.
13 Kant, *Critique of Judgment*, p. 172f.
14 "Facticity and project" (*Faktizität und Entwurf*): M. Heidegger's terminology in *Being and Time* (1927).
15 "*I am my body and have it at the same time*": the fundamental human situation, called by H. Plessner "*eccentricity.*"
16 W. van den Daele, *Mensch nach Mass*, Munich, Beck Verlag, 1985.
17 On this term, see Böhme and Schramm, *Soziale Naturwissenschaft*.

10 Body, nature, and art

Body and nature

The relationship of man with nature is determined in his own body. It is true that we have run into serious problems in our relationship with outer nature. The environment is burdened with toxic substances, the carbon dioxide content of the atmosphere increases, flood catastrophes are becoming a normal event, and even in Europe white patches spread on the maps: humans produce deserts. Yet there is no doubt: we are only hit by all this because we ourselves belong to nature, and do not only live off it. Drastically thrown back onto our own nature by what we are inflicting on outer nature, we become aware that we are ourselves nature.

The more civilization proceeded, the more nature became distant. It was this thing out there – beyond the town walls, on the other side of civilization, the wild matter not appropriated by work or shaped by technical methods. Through this distance we have forgotten how close nature is to us, and how we always do to ourselves what we do to nature. We have also kept the nature that we are ourselves at a distance, we have exploited it, manipulated it, and are today about to transform it. But what does it signify that we are nature ourselves? Do we know at all what it means to be nature?

Body: the nature that we are ourselves

What is the nature part of us seems to be clear. It is the body, which has been successfully studied by natural science and is put under the management of medical biotechnology. Does this mean being nature? Does science teach us anything about our nature, insofar as we are nature itself? The natural sciences treat our body like an object, i.e. like something external and before the eyes of someone else. The topic here is not how we can experience ourselves, how we are given to ourselves in our natural being. External experience is not self-experience. We can only get to know the nature that we are ourselves through self-experience: through physically feeling it. But who helps us to do this, gives us guidelines, orientation? Where does the language come from to articulate these experiences? If science doesn't, perhaps art will?

The closed and the open body

A wise man is reported to have said that the body was the actual subject of fine art. He must have been an idealist. Winckelmann and Schiller saw Greek sculptures as the expression of highest art, and that meant: the elevation of nature to an ideal rank. "It (is) humanity alone in which the Greeks comprise all beauty and perfection."[1]

Art as the imitation of nature, and the human body its most noble subject: does this mean that art has made human nature its topic? The beautiful human body, as presented by art, is closed. In spite of all the admiration, it remains outward, and the more it makes the intellect apparent (according to Hegel's demand), the more nature withdraws. The more, in art, the human appearance gains expression by glance, movement, gesture, and mimic, i.e. the more inner nature becomes visible in outer appearance, the more the inner being in its natural condition disappears. Schiller even said: "We perceive beauty everywhere where the substance is completely dominated ... by the form and ... by the forces of life."[2]

The beautiful body prevents the sight of nature. Accordingly, classical art remains on the surface in its representation of the human body. Certainty there is a counter-movement, and it does have consequences also in art. It is anatomy which pervades the beautiful illusion, wanting to disclose the human being in its natural state. What it finds is cleansed and arranged in categories of fibers, tissues, organs, and vessels. What was inward becomes outward, and at the end we find the "transparent human" in the Hygiene Museum in Dresden, and the plasticized preserved bodies in Volker von Hagen's collection. Anatomy, which was soon celebrated in the theater, became an object of fine art itself, and its knowledge has instructed sculptors since Leonardo da Vinci.

Closed in beauty or anatomically opened: the body in art does not appear as the nature that we are ourselves.

Naturally, things have not stopped at this constellation of beauty and anatomy. The representation of the human body in fine art has deeply changed in this century. In the wide range that has opened during this process, there will also be examples where the body as nature is depicted.

This does certainly not include impressionism such as Rodin's, stylization such as Giacometti's, or the breaking of visual taboos as done by Zoë Leonhard, but it does for instance include the anatomy of war by Otto Dix – as opposed to the anatomy of science. Further candidates would be: Francis Bacon, even though in his paintings he leaves anatomy and – perhaps in doing so – he approaches the perceptible body. Further, there is Petro Cabrita Reis who, on the *documenta 9*, mediated nothing to the visitors but the feeling of physical presence between two mighty walls. Finally, I would like to mention the sculptress Inge Maier-Buss, especially because she is not a true sculptress, i.e. she does not shape an opaque material from the outside, but she rather forms bodily features from transparent material and from the inside. What, however, is the actual point, what can be expected?

Bodily being in the world

That we are nature is first of all experienced in our natural state: we must eat and drink, must be born and die, must reproduce ourselves by procreation, are exposed to illnesses and pain. All this is present in fine art, and it hardly requires fine art to recall these facts. But in which form are they present? As symbols, gestures, and biographical narratives. But eating and drinking, for instance, are not merely cultural acts, nor merely the satisfaction of a want.

They are the execution of our nature. Paracelsus says that we live in the passage of the elements: breathing, the air passes through us; drinking, water runs through us; eating, earth trickles through us; and in intellectual education, fire nourishes us.

So, in breathing, eating, and drinking we can experience that we are pertinent to the elements. By analogy, the point in perception would be the experience of being pertinent to the things perceived, respectively to the media mediating them. Goethe has appropriately expressed this connection, for the visual field, in his well-known verse saying that, if our eye were not like the sun, how could we behold light? The basis for each perception is physical presence. Particularly in the times of telecommunication it is important to insist that, in order to perceive anything, one first of all has to be there. And because physical presence is the prerequisite for each perception, each single perception is based on the feeling of presence: we feel where we are in our inner awareness and emotional situation.

So the body is the primary sounding board enabling us to participate in the world. We are not in this world as specters or souls, but through physical presence. In an even more fundamental way than in our natural state we are nature because, in order to participate in the world, we must be physically present in it.

The body felt

Our nature is given to us in the manner of self-experience. We are this nature, our body, by feeling it. Which form it has, which dynamics, is only given to us in this feeling and remains hidden to the objectivizing view from outside. This sensing can be restraining, as in fear and pain; it can be widening, as in a sigh of relief; it can rhythmically fluctuate between restraint and widening, as in voluptuousness. The body is spatially not identical with the "mortal frame." We can go beyond the latter in sensing it, or be crushed together in it in an inevitable "Here."

The subsection of the sensed body does not correspond to the outer form, in trunk and limbs, but it is a manifold entity consisting of more or less connected, emerging, and disappearing bodily islands. In the dynamic act of bodily sensing we experience what we call our natural being. It is specified in the restraint of fear, in the uplift of joy, in the tension of feeling thirst and in the pressure of lust. In bodily sensing we can also experience our pertinence to the elements as they permeate us and carry us. Finally, we bodily feel that we are somewhere at all, our presence in surroundings.

Being a body – a task for art?

The modern human knows little of all this. As Günther Altner speaks of the dereliction of nature, this title would also include the nature that we are ourselves. The dereliction of nature as practiced by modern man is above all the dereliction of his body. What contemporary humans acknowledge as nature in themselves are the things presented to them by science: their organisms, their frames. They disregard bodily sensing by actively aiming at targets; personal feelings and convictions are subordinated to performance and output, and in perception the experience of self-presence is ignored while stating objects. Should art aspire at making the body as the nature that we are ourselves its topic, it would have to research a widely unknown field, i.e. the mediation of self-experience to modern man, a thing he has always suppressed. The point would be a liberation of the senses from their fixation on objects and signals – Hugo Kükelhaus has tried this in his experience field of senses. Even more elementary, the task would be to mediate bodily presence through installations and interactive surroundings.

This could also imply the nullification of the venue, for instance in the dissolution into light atmospheres, as happens in the works of James Turrell. Furthermore, the experience of bodily dispositions could be mediated by certain views and situations, as happening in erotic art as well as in the performances of Hermann Nitsch. Finally, there is the virtually unsolvable problem of representing the sensed body. It would demand a pictorial presentation of the human body, which would not only mediate appearances of movements and speak through confessions, but would enable the co-execution in own bodily feeling.[3]

But why all this? Because it is necessary. Because the relation of modern man with nature can only be revised when he experiences himself as nature. Art is research, many people say. Here is an unexplored field. Art teaches seeing, say others. Here it could teach feeling oneself. Art mediates experiences. So it should enable us to experience the living body as the nature that we are ourselves.

Notes

1 Friedrich Schiller, *Kallias oder über die Schönheit. Über Anmut und Würde*, Stuttgart, Reclam, 1971, p. 73.
2 Ibid., p. 40
3 Hermann Schmitz, *System der Philosophie*, II.1 "Der Leib," Bonn, Bouvier, 1965.

11 Nature as a subject

A comeback of nature

Nature as a subject had a comeback in the second part of the twentieth century, and it has intensified in the last decade to virtually becoming a fashion. It is high time to consider what this development has brought, and thus whether it may continue. In the first place, one must also question in what respect we are talking about a comeback of nature. This assumes that nature in art has lost – at least for the time being – its traditional and well-established position.

When one asks this question, one first encounters the astonishing and by no means understandable fact that art had a practically essential relationship to nature for the longest period of European cultural development. Superficially, one may thus express that art has always – that means since the cave paintings at Lascaux – allowed the forms of nature to be reproduced, that it was oriented toward nature as the paradigm of beauty, respectively, it owed its existence to the beautiful character of nature. However, this relationship has only been understood more deeply during the period where it became questionable, meaning since the eighteenth and the beginning of the nineteenth centuries. It was then expressively formulated that art is mimesis, an imitative representation of nature. Kant most impressively formulated this view in his *Kritik der Urteilskraft* (§ 45), where he writes that for art to be beautiful, it must be like nature. Simultaneously, this sentence is a transition to a greater depth in the relationship of art and nature, as found in the genius-aesthetic of the Romantic period. Thereafter, artistic endeavor itself is understood as nature's effect in and through the genius.

This background must be made clear to understand why one of the maxims with which modern art has appeared since Baudelaire was a withdrawal from nature. Making art autonomous was not only a social process, through which art was differentiated as a social subsystem with its own problem generation and its own controlling rules, it was also a turning away from nature as a paradigm of artistic creation. Making art autonomous was surmounting the maxim of mimesis: after which the artist saw himself as a creator, yes, as an inventor, and when as a genius, then not as a natural power, but as an aware and reflective artist. This did not mean that the forms of nature disappeared from art, but that they were understood to be material, freely available for the artist's use, as a stock of significances taken and used by art, making any reference to concrete nature no longer necessary. Moreover, one should not forget that this essential feature of modern art presented from the beginning a counter movement, which then practically fostered an explicit turning toward nature. Thus the Barbizon School, which was contemporaneous with the period of Baudelaire's creativity, is just as much concerned with the development of natural ornamentation in Art Nouveau as with the radical denial of any ornament, which Adolf Loos anticipated in the modernity of Bauhaus.

Nature: a problem of art

The relationship of art and nature in modern art is also quite ambivalent. To be aware of this is therefore the first criterion of quality by which works of art that take nature as their theme today must be measured. The wave of natural art in the last decade is undoubtedly motivated by the environmental issue. But this welcome engagement, with which the artists take issue, will lead to triviality if the artists do not realize that nature is essentially a problem of art, and not merely a matter of bringing the one over to the next; or respectively, that art is itself marked by the problem that one perceives as an environmental problem outside of art: namely alienation from nature. Nothing makes this clearer than the fact that the three or four genres of traditional art, which are characterized by the explicit selection of a central theme taken from nature, have been the cause of this alienation from nature. First of all, there are the "four elements" series by the copper engravers of Holland, and the emblematic of the sixteenth and seventeenth centuries; thus through their tableaux, so to speak,[1] the artists wished to regain the unity of the world, which at the beginning of the New Age was disappearing as artists' creations. Second, there is the art of still life, where nature as a natural element and as nature morte was lost as a vital correlation and was again made accessible to consumers as goods. Third, there is landscape painting – an expression of natural longing and a discovery of nature out there behind the development of the New Age individual and urban living. Fourth, there is the art of landscape gardening, and the way in which it developed from the middle of the eighteenth century, by arranging nature as a view or scene of effective harmony, just at the time when industrial exploitation came into being. All four spheres of fine art in this way show that they used nature explicitly as a centrally focused theme, which was no longer a given matter, and no longer an unnoticeable and integral part of art. Elementary art, emblematic, still life, landscape painting, and English garden art seem naive today. Contemporary, ecologically motivated art must be seen as a part of this tradition, but only in retrospect; it currently appears naive if there is a belief that nature can be alienated by focusing upon it. It does not escape this dialectic, unless it itself becomes a problem as art. Because – and that was another side of modern art – the bidding farewell to the mimesis idea, and thus nature as a paradigm of art, certainly forced art to define itself as an autonomous system, which it is. After Duchamp, Thierry de Duve writes in his book *Kant nach Duchamp* that the decisive question of aesthetics is no longer what is beautiful, but what is art? When it is simply a matter of nature as theme, it is essential today to simultaneously focus on what art is. Without this, art concerned with ecology is almost in danger of wandering off into the botanical, of becoming exhausted in a disgusted portrayal of waste, or to lose itself in a fog of conciliatory esotericism.

Nature: a problem for art

Wherever artists work with or in nature they are tempted to take part in the ecological discourse – even if it is only as a commentary in catalogs. But in this way the works of art are often given a mistaken legitimacy and loaded with superfluous interpretation. Entering into a relationship with the great tradition of European art about nature remains completely legitimate, if, while focusing on nature, it also takes into account the interrogative aspect of art, and thus abandons the naivety of these traditions. For example, artists work in this way within the domain of land art. It can certainly be seen as part of the tradition of English landscape gardening, to which human construction and sculpture also belonged. If Robert Smithson arranged stones in a remote area, however, the question of what is art is posed for those who experience an articulation of nature in them. Similarly, one may observe many examples of concrete natural lyricism. If Kant describes gardening as concrete landscape painting, one may consider many arrangements within nature today as concrete lyricism. They present a scene of nature, which, with a few further strokes, is brought to speak.[2] It is still clearer that nature is neither actively imitated nor made visually objective, and this takes place among those artists who directly allow nature to have an effect on their works of art, or who make art with nature. An example of this is Fridhelm Klein, who exposes his pictures to rain and ocean waves, or leaves them to fade through the weathering process. None of this is ecological art, but it presents the deeper problem that art shares with nature.

It is certainly to be applauded that artists become committed to the ecological movement; yes, one must say it is currently almost the only example of committed art to be found. Because the art of today, which has its own focus, and which is also primarily given credence by the art critics, is, however, in danger of becoming blind to the problems of the world in which we live. Much too occupied with the question of what is art, the artists avoid the question: why art? This question is much too often not posed, and it all too easily makes art a luxury phenomenon, namely, a market-dominated playground. That art is a serious matter is felt everywhere today in countries in which politically repressive systems rule. From the viewpoint that, if ecologically engaged art takes its social role seriously, it is by all means legitimate. But now the question of quality is posed all the more urgently, the more art serves another cause than the intention of simply being art. Artists could thus have transcended the annoying question of what is art, by merging into life, as demanded by the avant-garde, or, as Joseph Beuys formulated, follow a broader art term. But when is ecologically oriented art good? The answer must be: ecologically oriented art is good when it confronts the actual problem, in one way or another engaging in really offering a contribution to solving ecological problems. The question of what actual problems we have with nature, and to what degree art can somehow contribute to solving them, however, must be posed *before* the artistic work commences. All too often curators invite artists to exhibitions that in some way focus on the theme of art and nature, without a theoretical preamble. The contribution of art historians and philosophers are first called upon when the works of art are already selected. It may well be that artists are often particularly sensitive contemporaries, but they understand themselves and we understand them wrongly if they are highly regarded as superior philosophers or natural scientists. Thus, the contributions of artists engaged in the ecological discourse are unfortunately often at the level of an average newspaper reader and pub philosopher.

The problem of ecology is a crisis of the socially organized relationship with nature. For that reason, it must be resolved primarily on the political, economic, legal, and technological levels. Art can contribute almost nothing on these levels, unless it is in the insubstantial form of a creation of awareness. But that awareness, which is deeply disturbed in our relationship to nature, is no longer required. Today everything is a matter of solutions. Art can genuinely become engaged with the means at its disposal in three areas, which are to be labeled under these key terms: natural oblivion, the failure to master nature, and a humane return to a natural appropriation.

The problem of ecology is very much a problem of the relationship humans have with themselves. The distanced and instrumental attitude taken by humans toward external nature is also taken on as a part of nature, which they themselves are. When one wishes to speak here of natural oblivion, it is primarily a matter of humans who have forgotten their affiliation with nature. Through the ecological problem, which means through what we do to external nature, we are unmistakably reminded that we will eventually suffer ourselves. In the resulting task of integrating nature as an essential, intrinsic part of the way that humans see themselves, art could really contribute something. It is a matter of regaining the sensual forms of existence, of the rediscovery of the body and the rehabilitation of physical presence. This is at the same time the point where the contribution of art to deep ecology also poses the question of what is art in a work. One possible answer to this question is the exercise of alternative means of perception, and for art that means the actualization of the aesthetic view. Sensually pedagogic art comprises the possibilities given in art from Hugo von Kückelhaus to art as a performance of sensual experience, as in James Turrell's work, in which the experience and regaining of one's own nature is made possible.

As projects of the New Age, natural science, and technology were approached to manage the practical emancipation of humans from nature. But with their expenses and side effects they have led humanity into an even greater dependency. Human life has perhaps been extended on average, and the number of people on the planet has monstrously increased, but the average statistics of human illness have not improved, and the number of people wiped out by hunger and pestilence has also increased, together with the destruction caused by natural catastrophes. Toxic materials are dispersed throughout every corner of the world, fertile soil is lost at a rapid rate through erosion and over-fertilization, and above all: the human power of destruction through natural science and technology has by far grown quicker than our measures for the preservation of life. In short: the project for mastering nature has failed. Faced with these facts, the slight degree of real confrontation with this theme in art is astonishing, and above all, that nature in the sense of natural catastrophes appears to such a slight degree. Art observers hardly feel the terror of nature in their bones as expressed in the fire installation by Marie-Jo Lafontaine, or in the chain messages by Paul Garrin at the Mediale in Hamburg. Certainly, there is also criticism of science and technology, but it is often really nothing more than an aesthetic flirt or an inquisitive mental fairy tale in the mode of science fiction. The classic environmental art has always presented an alternative to natural science and to a technological relationship to art, and in the view of Alexander Gottlieb Baumgarten these were even to be read as an alternative knowledge of nature. Here, too, art would have a task today. But what artistic research there is about nature hardly rises above the private reading of tracks and nostalgic amateur science. It is at its worst when this research becomes private mythology, as with Joseph Beuys. This science lacks inter-subjectivity and ideas in which the other view can be maintained.

According to Baumgarten, art as a sensory acknowledgment of nature shares the alienating attitude that people have toward nature with natural science. Another awareness of nature will only be possible when people include themselves as subjects of this awareness as natural beings. Other than showing what nature presents, art could at least prepare an investigation of it as theme. Our historic place is thus defined, in that we are not on the threshold of an environmental catastrophe, but in the middle of it. Even in a climatically and economically moderate Europe, the desert advances, and the nature surrounding us is to a large extent already destroyed. The natural cycles are shattered, biodiversity is significantly reduced, and above all there are already gigantic areas that, according to human estimation, are to be described as deserts: areas of brown coal waste, uranium-contaminated mountain ranges, fallow fields of industry and traffic. To regain these areas, which means to bring them back into order, in which an existence worthy of humanity is made possible, is *here* a completely endless – meaning economically unviable – task. What one calls re-naturalization and re-cultivation is so extensive for enterprises that they are only possible in a small part. But where they can be undertaken, the ecological and economical viewpoints are too slight to reconvert the desert to a humane level of nature. It is here that art has its real task – in the reclaiming of destroyed nature. But orientation about the great, classical, cultivated,

European landscapes and parklands is insufficient because the framework conditions of nature to be designed today have changed completely. Classic nature was the nature out there around the human settlements and the sphere of civilization. Nature today is an intermediate nature, the space between urban agglomerates, technological infrastructure, and traffic thoroughfares. New aesthetic concepts must be drawn up in this respect. It is here that even artists may assume a leading role and, like the Harrisons,[3] organize social processes for a humane reappropriation of nature. The work of Hermann Prigann – *Terra nova* – should be mentioned in the same way.

The time in which protest was an art is past. Art can make a contribution to mastering the environmental problem:

- by allowing people to experience their own nature;
- by offering counteraction to the technological acquisition of nature and by posing the question of what nature is; and
- and by initiating and organizing the humanizing of already destroyed nature.

Notes

1 Klaus Popitz, *Die Darstellung der Elemente in der niederländischen Graphik von 1565 bis 1630*, Munich, Diss. Phil., 1965.
2 The Schüberg-projects have to be mentioned here (*Nunatak. Projekt: Schüberg. Die Natur sprechen lassen, Bad Oldesloe: Kulturstiftung*, Stormann, 1983, and *Natur- und Kunstsymposion Hier und Da, Naturschutzgebiet Schüberg*, ammerbeck bei Hamburg 1992), as well as a few works of the series *NaturArte* by Werner Henkel.
3 Helen Mayer Harrison and Newton Harrison.

Part III

Architecture

12 The atmosphere of a city[1]

Odors

In the past, the metro in Paris had a very distinctive odor. Someone could have brought me to Paris while I was sleeping and I would have immediately recognized where I was, just by this odor. Today I would give a lot if someone could get me a little bottle of that scent. I would sniff it and from the scent I would smell the Paris of that time, just as Marcel Proust could smell his Combray in a *madeleine* cake. Paris has changed; it is much more technologically developed and cleaner, and today you would probably have to recognize it by something other than its odor. Perhaps my Paris odor was the last whiff of those miasmata which, according to Alain Corbin's wonderful presentation in his *The Foul and the Fragrant,*[2] have been forced out of the city in a number of deodorization campaigns by the newly sensitized citizens, since the beginning of the nineteenth century.[3] But perhaps today it is just that younger people recognize their Paris by different odors, whereas I, in a nostalgic mood, refuse to perceive them. That cities, districts, neighborhoods, and landscapes have their odors is still true today – in spite of sewage systems, ventilation, or deodorization. So, for example, your nose can still tell you whether you are in East or West Berlin. This is due no longer to the fumes from two-stroke engines, but to the use of brown-coal briquettes in the eastern part of the city. And we know we are somewhere else thanks to the soil, when it is damp, or to the stones, or to particular trees that grow in the city, or because of the fact that you can smell the sea, because of the gasoline used, the means of transportation generally, and of course because of the people themselves, their lifestyles, and eating habits.

Odors are an essential element of the atmosphere of a city, perhaps even the most essential, for odors are, like almost no other sensible phenomenon, atmospheric: "Expelled indeterminately into the distance,"[4] they envelop, cannot be avoided; they are that quality of a surroundings which most intensely allows us to sense through our disposition (*Befinden*) *where* we are. Odors enable us to identify places and to identify ourselves with places.

Hence, it is not surprising that the first scientific book that dealt with the topic of atmosphere was essentially a book about odors. I am referring to

Hubert Tellenbach's book *Geschmack und Atmosphäre*.[5] As a psychiatrist, Tellenbach was primarily interested in the disorders of "odor atmospheres." Even more evident in his approach is the fact that trust between people is founded on the "atmospherical."[6] Using the analogy of nest odor, among animals, he sees the intimacy of the space occupied by the family and one's native place as constituted by the odor atmosphere.[7]

It is difficult to talk about our experiences with odors, especially in an academic discourse. That is why Tellenbach referred to literature, especially Dostoyevsky. To clarify what has been said so far, I would like to refer to another Russian author, Nikolai Gogol, in particular to the passage in his novel *Dead Souls*[8] where he describes Chichikov's serf Petrushka:

> He slept without undressing, just as he was, in the same coat, and he always carried about with him a peculiar atmosphere, a smell of his own, which recalled the stale air of a stuffy room, so that it was quite enough for him to fix up his bed somewhere, even in a room which had not been lived in before, and to bring his overcoat and belongings there, for it to appear as if people had been living there for a dozen years.[9]

Here we see the phenomenon pinpointed with a beautiful ironic twist: whereas rooms, because they are lived in, ordinarily have a specific odor peculiar to them, one that makes them snug and in which we can feel at home, the dumb but crafty servant Petrushka carries his own (for Chichikov, quite penetrating) atmosphere around with him, so that he can lie down anywhere and feel immediately at home.

The odor of a city – perhaps it is indeed this atmospheric element whose neglect or even whose banishment makes our cities so "inhospitable," as described by Mitscherlich[10] in his well-known pamphlet. A city without an odor is like a person without a character.

As matters stand

The question concerning the *other dimension* has run through literature for as long as there has been modern city development and city planning. Somehow one senses that there must be something else besides city zoning and traffic, besides functional division and organization. In his then very influential book *Der Städtebau nach Künstlerischen Grundsätzen*,[11] Camillo Sitte called this "something else" the *aesthetical*. Gordon Cullen speaks[12] of the *townscape*. Kevin Lynch calls it the *image* of the city.[13] A city's image is an expression of its self-presentation, what impression it makes, the atmosphere it radiates. With reference to Alexander von Humboldt's use of the phrase "total impression," the term "townscape" can mean indeed much more than what we simply see. Nonetheless, a limiting of the investigative perspective to the visual or even to the geometrical is evident among the authors mentioned. This becomes clear as soon as the analytical categories are scrutinized. In Lynch: paths, borderlines, districts, focal points, signs; in Cullen: serial, vision, place, and content. As well as color, texture, scale, and style, "content" does indeed include character, personality, and uniqueness. But in Cullen we read: "For it is almost entirely through vision that the environment is apprehended."[14]

We have to turn to another, older writer if we want to learn something about the atmosphere of a city, not from *belles-lettres* but from the literature on aesthetics: August Endell. Against the polemic of his time, popular among the youth and inspired by the Impressionists' pictorial conquest of the metropolis and technology, this Art Nouveau artist and theorist wrote a book titled *Die Schönheit der grossen Stadt*.[15] Because of its effusive and unsystematic presentation, this book may serve today merely to be exploited as a source of ideas. But Endell did point out many phenomena which we can now gather under the concept of "atmosphere." To mention a few instances: the city as nature,[16] the city of sounds,[17] the city as a landscape,[18] the day's veils (fog, air, rain, dusk), the street as a living being,[19] people as nature.[20] What Endell pointed out was the fact that nature unfolds in the city too, that the city has its own characteristic life and that the beauty, the poetic quality of a city can transpire entirely at odds with its architecture and planning. I shall return to Endell; but to do him honor at this point, too, I will quote a brief, characteristic description from him:

> It was during the hot summer, somewhere on the northside by the belt line, where the rail tracks on the bridge were not carefully laid in consideration of the ears of those living nearby on a muffling sand bed, but put directly on the construction, without cushioning, rattly. Under such a bridge there stood a cart with timber beams, two heavy horses in front, their mighty heads drooped, tired. They stood to one side of the street in front of a yellow brick wall, and their standing there made the underpass seem bigger, wider. On the other side – to make the Space more vivid – there stood two children. Outside, the sun was brooding in a stifling haze, the brightness seemed to demarcate, like a transparent coat, the space to the fore and to the rear, and this space was filled with blue shadows. But through the holes in the rail construction, just like through the branches of a tree, thousands of individual sun rays beamed into this cool shade, onto the dusty street, onto the children, onto the yellow timber, and onto the silent, colossal horses.[21]

Commenting, then, on what he senses here as a whole, Endell writes: "It is the life of space which gives form and color here, as in similar cases, such a strong, meaningful foundation."[22] We call it atmosphere.

The concept of atmosphere

It is not at all unusual to speak of the "atmosphere of a city." We find the expression in everyday speech and in writing, in advertising material for cities and in the travel supplements of newspapers. For this use of the term in ordinary language, two things apply: first, atmosphere is mentioned, as a rule, from or for the perspective of the stranger; second, this is an attempt to identify something characteristic of a city. When we speak of atmosphere as something which is experienced by strangers to the city, then this expressly implies *not* saying that we mean the city from the tourist's perspective. Rather, what is meant by atmosphere is that which is commonplace and self-evident for the inhabitants and which is constantly produced by the locals through their lives, but which is noticed first by the stranger as a characteristic. This is why the *atmosphere* of a city is not the same as its *image*. The image of a city is the consciously projected self-portrait, and the sum of its advantages, that an outsider might enjoy. Moreover – to come to the second point – by "the atmosphere of a city" we understand something characteristic, that is, something peculiar to the city, what makes it individual and therefore cannot be communicated in general concepts. That does not, however, mean that we cannot talk about the atmosphere of a city; we shall see that that is indeed possible; rather, what it means is that atmosphere is something that has to be *sensed* in order to understand what is really at stake when we talk about atmosphere. The atmosphere of a city is precisely the way life unfolds within the city.

Working the "life worldly" meaning of atmosphere into a concept in aesthetic theory is first of all advantageous to aesthetic theory itself; in this case, to the aesthetics of the city.[23] Introducing the concept liberates this aesthetics from being restricted to the visual and symbolic. Anything that cannot be grasped in structures is shifted to "meanings." Thus, for instance, there is talk of *Meaning in Western Architecture*[24] or of *The Language of Post-Modern Architecture*.[25] Here, however, one is just following the trend set by semiotics, and one fails to realize that the age of representation has long come to an end.[26] To put it another way: the multicultural world of our large cities does indeed contain more and more universally understood pictograms, but it no longer has a symbolism understood by the community as a whole. And that means that which appeals to us in a city cannot be construed as language; instead, it enters our disposition as a touching character (*Anmutungscharakter*).

This brings us to the second advantage for an aesthetics of the city which makes use of the concept of atmosphere. In such an aesthetics it is a matter not of how a city should be judged in terms of aesthetic or cultural aspects, but of how we feel in this city. Here we are taking a decisive step toward including what has been, somewhat inappropriately, called the "subjective factor." Of course, we only ever sense an atmosphere in our own perception, but on the other hand we sense it as something which radiates from another person, from things, or from the surroundings. To that extent, it is something subjective which can be shared with others and about which an understanding can be reached. In studying atmospheres, it is a question of how we feel in surroundings of a particular quality, that is, how we sense these qualities in our own disposition. We can reach agreement on such dispositions by pointing out their character. An atmosphere can be relaxed or oppressive; it can be businesslike, jovial, or festive. Our languages have countless expressions for characterizing atmospheres, whereby we can distinguish two main groups. First, our languages have terms for synesthetic character, that which is sensed primarily in a modification of our bodily disposition. Second, there are terms for social character, that which is coined by social conventions. Examples of the latter are elegant, *petit bourgeois*, meager. Using atmospheric characteristics to help analyze urban environments would be, historically speaking, an extension of what Hirschfeld[27] introduced in his descriptions of park scenes. It would aim at determining urban environments with respect to the "feeling of life" (*Lebensgefühl*) for those who live in them or who visit them, and would include identifying the causes of possible pathologies.

The third advantage in the concept of atmosphere is found on the objective side. We cannot study atmospheres solely from the side of the subject, that is, by exposing ourselves to them; they can also be studied from the side of the object, that is, from the side of the agencies (*Instanzen*) by means of which they are created. The paradigm for this perspective is provided by stage design. The general aim of stage design is to create an atmosphere with the help of lights, music, sound, spatial constellations, and the use of characteristic objects. The paradigm of stage design falls short of the case of city planning, inasmuch as the atmosphere in the latter is created not for the outside observer but for the actors, as it were; that is, for the participants in urban life, who together produce the urban atmosphere through their own activities. Years ago, already, Werner Durth[28] pointed out the dangers involved in understanding city planning as stage-managing. They consist in what Walter Benjamin criticized as being an aestheticization of politics, which is, "allowing the masses to express themselves (but by no means allowing them to enjoy their rights)."[29] But the paradigm of stage design offers the advantage of providing a wide range of categories and instruments according to which atmospheres can be determined from the side of their creation. The question concerning the *generators* of atmospheres broadens the perspectives and possibilities of city planning considerably – and, it is hoped, its responsibility too.

Generators of atmosphere

With regard to the domains or dimensions of the *generators* of atmospheres I would like to concentrate, in my examples, on those which have so far enjoyed little attention in the literature on city aesthetics and city planning, or have found no place there at all; I am referring to the domains of the acoustic, and of lifeforms. Of course, the dimension of the visual, as well as spatial structures and visible forms, incorporate generators of atmosphere, or they can be interpreted in this way. Correspondingly, it would not be a question of what form, for instance, a building has, or how a city is structured, but of what it radiates because of these features, or in what way it codetermines the dispositions of the inhabitants. Models based on such an interpretation can indeed be found in classical literature; for instance, where the question of *orientation* is central in Kevin Lynch's writings. Similarly, Cullen occasionally speaks of the significance of the geometric structure he identifies for "the position of our body."[30] To be more precise, here it would be a question not simply of relative positions, but of how we are bodily disposed in such and such a structured space. It does make a difference whether we go through narrow lanes or across wide esplanades; whether winding, hilly streets, or long straight ones are characteristic of a city; whether among the skyscrapers we suddenly come across a little church or, on leaving a lane, we find ourselves on an extensive square. Spatial structures and constellations are not merely seen and assessed, they are also sensed by the body. In this respect, existing studies would have to be reinterpreted. Much the same applies to a domain that has so far been dominated by the discourse on signs, the domain that I would like to call the *historical depth* of a city. It is of course an enormous pleasure for the learned to be able to *decipher* a city, that is, when its history becomes transparent through stylistic features, heraldry, epigraphs, materials employed. But these abilities can no longer be assumed to pertain to the average citizen, and historical information frequently denies the guided tourist the possibility of experiencing anything of the city. But "being old" or "having grown over time" are city qualities which do not manifest themselves solely in signs; rather, they are touching qualities (*Anmutungscharakter*) that are *sensed*. These could, perhaps, be the same qualities that can be read as signs, the ancient material or the ancient line of architecture, for instance, but sometimes they could be completely different qualities. So, for example, we sense the historical depth of the city of Lübeck – or Maastricht – by the fact that the churches are, as it were, *rooted*; they rise, so to speak, from the ground, like trees. But we are also familiar with the disillusioning effect of the cleaning of the stained-glass windows in Chartres cathedral. That is to say, inversely, that it is quite possible that a historical sign which suggests a distant epoch is precisely *not* the reliable clue for divining the historical depth of a building. In this sense, faithful reconstruction can be as counterproductive as the removal of ivy from an old tower. The dimension of historical depth or the atmosphere of an organically developed city is, however, of major importance for the inhabitants' sense of feeling sheltered and at home.

In city planning, the acoustic dimension has so far been dealt with almost exclusively in quantitative terms, that is, with regard to noise pollution and its avoidance. The *character* of sounds has, on the other hand, almost never been a topic. The major exception here, once again, is Endell, who, under the heading "The City of Sounds,"[31] joyously pursues the manifold voices in the city. However, I would like to let another witness – one who lived at the same time and in the same city – have his say: Victor Klemperer, in his *Curriculum Vitae*;[32] and I choose to quote him for the reason that in his text it is clear that the manner in which sounds are experienced in the city not only depends on decibels, but also on their character:

> I had found a room in Dennewitzstrasse. The house was very working class and not very hygienic. The windows opened onto the nearby railway

embankment; the expanse of the rail network of the Anhalt and Potsdam stations unfolded there; day and night I was surrounded by the images, the many colored lights, the rumbling and whistling, the cries and horn signals of the immense rolling stock. That touched me again and again like a wonderful promise.[33]

The worldwide project "Soundscape" has done excellent preparatory work in researching into the city as an acoustic space. In this project it was composers and sound engineers, above all, who were concerned not only with the recording and composition of natural sounds but also with the acoustic profile of cities. One of the studies, for example, by the founder of *Soundscape*, Murray Schafer, dealt with the city of Vancouver. Soundscape researchers distinguished between an acoustic background (which of course changes in the course of the day) and characteristic sound events. It was discovered that, to communicate the character of a soundscape, it is not sufficient just to present a recording; it is necessary to condense and above all to compose. An excellent example of a composition which conveys the sound atmosphere of a city is H. U. Werner's *Chicago*.[34]

In the meantime, city planners have also taken up the topic. Thus, Pascal Amphoux researched *La qualité sonore des espaces publics européens*.[35] His particular perspective here was determined by the hypothesis that the sound atmosphere of the city depends on national culture as well as the various forms of life. For this reason, he studied, in the case of Switzerland, the sound atmospheres of Lausanne, for the French region, Locarno, for the Italian region, and Zurich for the German-speaking part. Meanwhile, studies conducted by the Darmstadt city planner Barbara Boczek show that it is possible to identify differences in the sound atmospheres of pedestrian zones in German cities.

The question of sound atmospheres is directly related to the dimension of lifestyles, understood as generators of urban atmospheres. In terms of street noise, it makes a difference whether it is customary for people to honk their horns or not, what make of car they drive, whether radio music can be heard through their open windows, whether the names of goods are shouted out, or "alluring" music comes from the boutiques – these are just some aspects: through their lifestyles, the inhabitants of the city are also, always, producers of its atmosphere. Again, a literary witness for this, and once again it is Victor Klemperer:

> In spite of the work, which for many daylight hours encapsulated me as if in an empty room, I remained constantly conscious of the Parisian atmosphere. This was already taken care of by the cheerful hustle and bustle in the small restaurants during lunch and dinner time, where at least one fat cat carelessly wandered between the feet of the guests and waiters on the sandy floor, one which probably caught a mouse occasionally and ate it quietly; and then the coffee counter, where you

sometimes saw coachmen standing, the whips with the short handle and long strand thrown over their shoulders. Indeed, it seemed that my receptiveness for Parisian life grew with the intensity of my eagerness to work and with the shortage of time.[36]

Of course, I have chosen this text because it is also evidence of an early use of the term "atmosphere" to designate the total impression that is regarded as characteristic of a city, in this case Paris. Here, Klemperer condenses this impression and ties it to a particular scene, that of the small Parisian cafés and restaurants. In so doing, he refers to particular ingredients, like the sand strewn on the floor; he points out the nonchalance of the restaurant, in allowing cats inside; he sketches the fleeting presences of different types of people, and their occupations (still identifiable at that time). One could of course add to this something from the novels of Zola, Proust, and Döblin, as well as from Walter Benjamin's unfinished study on the arcades of Paris.

The rule that we have to *sense* atmospheres applies especially to the atmospheres created by particular ways of living. In order to really get to know them, we ourselves have to enter into them fully, as it were. That is why film enjoys a certain advantage in communicating them. In fact, in film use is often made of a particular urban atmosphere in order to generate the right atmosphere for a dramatic scene. A city atmosphere is very rarely, in itself, the subject of a film, Wim Wenders' offering some of the few examples (Berlin, Tokyo, Lisbon).[37]

We can study the contribution of lifestyles to the development of the urban atmosphere, but they are not an object of planning. However, we can think about which measures in city development promote or prevent particular ways of living. Here we find a link with the critical studies of psychologists and sociologists on the modern city.

Conclusion

The concept of postmodernity comes from architectural theory. It served to indicate that something has come to an end in the development of architecture, namely modernism, which, especially for architecture, meant the domination of functionalism. For city planning, according to the Charter of Athens, this has meant the separating of the city's basic functions: residence, recreation, transportation, and work. It actually led to the desolation of inner cities, the intensification of commuter traffic, the development of satellite towns, and city sprawl. The failure of this concept was made clear to architects and city planners by other professionals, by psychologists, and sociologists. But innovation could not, strictly speaking, come from this side; at best they could offer suggestions for compensatory measures to attenuate the mental and social harm done by modern city planning – psychologists and sociologists have *only* the subjective factor at their disposal. But for architects and city planners, the question is what concrete measures can change or develop cities in such a way that the mental and social harm criticized by psychologists and sociologists does not occur in the first place, and that life in the cities becomes tenable, or even attractive. The necessary concepts and objectives have been specified, at least, in the discourse between architects and city planners, and their critics, and they are: urbanization, residential environment, identification possibilities, city image, staging the everyday world, or more generally and correspondingly more vaguely, "aesthetics." However, really new, integrative concepts have not yet emerged from architecture and city planning, as a result. Wolfgang Welsch's proposal[38] to grasp the essence of postmodernity as plurality, does not fill it with content because it is no longer comprehended as a radical revision of modernity. Postmodernity is then just understood as a continuation of modernity, though with other means and at another level. The classical dichotomies are retained – subject and object, culture and nature, rationality and feeling – and, correspondingly, so are the familiar suppressions and disavowals.

Here, the concept of atmosphere could first of all, and at least, change perception. It directs attention to the relation between the qualities of surroundings and dispositions.[39] The atmosphere of a city is the subjective experience of urban reality which is shared by its people. They experience atmosphere as something objective, as a quality of the city. And it is indeed the case that, by analyzing the generators of atmospheres from the point of view of the object (i.e. through city planning), we can bring about the conditions in which atmospheres of a particular character can develop. The dimensions and possibilities for action that city planning has are thereby extended; but also, necessarily, is its attitude, for in the domain of atmospheres action does not always simply mean doing, it also means forbearance.

Notes

1 I would like to thank Barbara Boczek, Werner Durth, Mickael Hauskeller, and Dieter Hoffmann-Axthelm for suggestions and information. The text was translated into English by John Farrell.
2 Alain Corbin, *The Foul and the Fragrant: Odour and the French Social Imagination*, trans. Miriam L. Kochan, Cambridge, MA, Harvard University Press, 1988.
3 See also Ivan Illich, *H₂O and the Waters of Forgetfulness: Reflections on the Historicity of Stuff*, Dallas, Dallas Institute Publications, 1985.
4 A general determination of feelings as atmospheres according to Schmitz. Hermann Schmitz, *Der Gefühlsraum*. Vol. III, "System der Philosophie," Bonn, Bouvier, 1969.
5 Hubert Tellenbach, *Geschmack und Atmosphäre*, Salzburg, Otto Müller, 1968.
6 Ibid.
7 Ibid.
8 Nikolai Gogol, *Dead Souls*, trans. David Magarshack. Harmondsworth, Penguin 1961.
9 Ibid., p. 30.
10 Alexander Mitscherlich, *Die Unwirtlichkeit unserer Städte. Anstiftung zum Unfrieden*, Frankfurt/M., Suhrkamp, 1965.

11 Camillo Sitte, *Der Städtebau nach künstlerischen Grundsätzen*, Braunschweig, Wiesbaden, Birkhäuser, 2002 [1883].
12 Gordon Cullen, *The Concise Townscape.* 5th edn, London, Architectural Press, 1968.
13 Kevin Lynch, *The Image of the City.* Cambridge, MA, MIT Press, 1960.
14 Cullen, *The Concise Townscape*, p. 8.
15 August Endell, *Die Schönheit der großen Stadt*, Stuttgart, Strecker und Schröder, 1908.
16 Ibid., p. 30.
17 Ibid., p. 31.
18 Ibid., p. 33.
19 Ibid., p. 65.
20 Ibid., p. 66.
21 Ibid., p. 74.
22 Ibid., p. 75.
23 Gernot Böhme, *Atmosphäre. Essays zur neuen Ästhetik*, 2nd edn, Frankfurt/M., Suhrkamp, 1997.
24 Christian Norberg-Schulz, *Meaning in Western Architecture*, 3rd edn, New York, Praeger Publications, 1977.
25 Charles Jencks, *The Language of Post-Modern Architecture*, 4th edn, London, Academy Editions, 1984.
26 Harry Redner, *A New Science of Representation*, Boulder, CO, Westview, 1994.
27 C. C. L Hirschfeld, *Theorie der Gartenkunst*, 5 vols., Leipzig, M. G. Weidmanns Erben und Reich 1779–85.
28 Werner Durth, *Die Inszenierung der Alltagswelt. Zur Kritik der Stadtgestaltung.* 2nd edn, Braunschweig/Wiesbaden, Vieweg, 1988.
29 Walter Benjamin, "The Work of Art in the Age of Mechanical Reproduction," in Walter Benjamin, *Illuminations*, trans. Harry Zohn, London, Fontana, 1973, p. 243.
30 Cullen, *The Concise Townscape*, p. 9.
31 August Endell, *Die Schönheit der großen Stadt*, Stuttgart, Strecker und Schröder, 1908, pp. 31–3.
32 Victor Klemperer, *Curriculum Vitae. Erinnerungen 1881–1918*, 2 vols., Berlin, Aufbau, 1996.
33 Klemperer, *Curriculum Vitae*, vol. 1, p. 401.
34 Hans U. Werner, Uli Tobinsky, "Chicago City on the Move," in Detlev Ipsen et al. (eds.), *Klang-Wege*, Kassel, Gesamthochschule, 1995.
35 Pascal Amphoux, *Aux écoutes de la ville. La qualité sonore des espaces publics européens – Méthode d'analyse comparative. Enquête sur trois villes suisses*, Zurich, Schweizerischer Nationalfond, 1995.
36 Klemperer, *Curriculum Vitae*, vol. 2, p. 52.
37 Wim Wenders, *Tokyo-Ga*, 1984/85; *Wings of Desire*, 1987; *Far Away, So Close*, 1993; *Lisbon Story*, 1994.
38 Wolfgang Welsch, *Unsere postmoderne Moderne*, 3rd edn, Weinheim, VCH, 1991.
39 Gernot Böhme, *Für eine Ökologische Naturästhetik*, 3rd edn, Frankfurt/M., Suhrkamp, 1993.

13 Atmosphere as the subject matter of architecture

The difficulty of talking about architecture

Talking about art is always difficult, as demonstrated not only by the sorry state of catalog essays and art criticism but also by the fact that two whole disciplines must be enlisted in that discourse: aesthetics and art history, whose social function is to overcome the speechlessness of beholders by way of professionals who furnish the relevant categories of discussion. Talking about architecture seems an even more troubled practice, at least when architecture is treated as art – that is, when a building is not only functional but has a *surplus* of some kind, as Adorno puts it, providing speechlessness its specific occasion. It is especially difficult to talk about architecture by relying upon classical aesthetics, according to which buildings should, at one and the same time, be functional and, as works of art, *be functional without possessing a function*. This contradiction, or rather the dialectic that it generates, was boldly exploited by Hegel to posit a three-phase history of architecture since antiquity.[1] Presuming that it is correct to argue that the architect – quite unlike artists of other genres – creates a work of art that must couple the artistic with the useful, then one is naturally tempted to interpret the artistic character of architecture by borrowing from other arts and drawing comparisons with sculpture, painting, literature, and music. A building is like a sculpture; the architect proceeds in his sketches like this or that painter; a space speaks a poetic language; a construction has a structure like a Bach fugue. Such talk is, of course, meant as applause, and yet one wonders if it is not simply the product of discomfiture or even condescension. Does architecture really have nothing to call its own?

The affinity and the exchange between the arts are no doubt both important and noteworthy. But the disposition to speak about architecture in a way derived from the other arts is not only detrimental to the reception of architecture, because it obscures its own genuine concerns in a fog of metaphors; it is also a danger to architects. It leads them astray with a borrowed self-image; it beguiles them into basing their work on an understanding of the artistic that has been lifted from other arts. Having come full circle, the discourse now coheres: one architect designs his buildings like sculptures, another tries a painterly approach, a third wants buildings to be like texts, and a fourth like music. And why not? Why shouldn't the drawing of such relationships be a fruitful heuristic procedure for the architect and an enlightening metaphor for the beholder? They are. But they could also be excuses – a means of sidestepping what really counts in architecture.

So, what does really count? If we briefly review the basic implications of the comparison with other arts – form and content, expression, meaning, harmony – then sculpture seems to be the closest to architecture. Don't the two fields, inasmuch as they both shape matter, work in the domain of the visible? At which point the architect, by working for visibility and treating design as lending form to mass, has already succumbed to the seduction of the arts. But, then, is seeing really the truest means of perceiving architecture? Do we not feel it even more? And what does architecture actually shape – matter or should we say space?

The perception of architecture

Hegel, who classifies the arts in terms of the senses, assigns architecture to the visual arts without giving the matter much thought – possibly influenced by a partiality to vision inherited from the Greeks. Today there are entirely different reasons for classifying architecture as a visual art. This view is now based largely on self-representation or rather the presentation of works of architecture. Long before construction starts, the presentation of architectural projects in drawings, models, and, more recently, computer simulations and animations has become essential, for competitions and for clients. And afterwards, once the project has been finalized and the building completed, the representation of the work in photographs has become just as important as, if not more important than, the building itself. The skill with which architects are presented in trade journals, catalogs, newspapers, and brochures is vital to establishing a reputation and depends upon the successful photographic representation of their works. After all, how many people can travel all over the world to get an impression, *in natura*, of the works produced by the luminaries of architecture? It is little wonder, then, that thoughts of later photographic rendition already enter into the design stage of an architectural project.

We have hereby named the third factor that determines architectural creation: architecture must not only be useful and functional, it must also be a work of art – and it must be paid for, it must have a marketable appeal. That means advertising and branding. It also means staging the architecture, which explains why architecture today has a tendency to stage its makers as well.

And yet, if architecture really does consist essentially of the design of space, then it does not belong to the visual arts. You cannot see space. One is tempted to argue the viability of this statement on the basis of the inadequacy of perspectival representation, but that would involve jumping to the conclusion that what one actually sees (namely a picture) is flat – which in turn leads to the banal conclusion that no amount of illusion can adequately reduce three dimensions to two. The fallacy lies in the fact that we conventionally consider the camera a model of seeing with the eyes – with one eye! But vision obviously involves two eyes, and no amount of technology has ever succeeded in replicating what it shows us without recourse to the eyes.

So we see space after all, because we see with two eyes. But what do we actually see? And what is the feat of binocular vision? Once again, we tend to define its achievement in technical terms, namely, on the model of the binocular telemeter, which determines the distance of objects by calculating from a fixed base with reference to the two end points of that base. And this is the way vision estimates distances as well. But there is another important effect of seeing – one that, incidentally, radically contradicts perspectival vision. The art historian Ernst Gombrich was quite justified in treating the superimposition of one object on another as a central issue of perspective: painting in per-spective means painting so that nothing appears that cannot be seen by an eye fixed on a particular point. But binocular vision undermines that very principle: one can see around obstacles to a certain extent and the *fuzziness* that is thereby generated invests things with the quality of floating in space. Add to this the movement of the eyes: through the constant change in perspective, things become quasi experimentally displaced. As paradoxical as it may seem, the impression that things are in space is conveyed by the very fact that their location is indeterminate.

Apparently, a distinction must be made between the physicality of things and their existence in space, that is, their ability to establish space through form or arrangement. Perspective is clearly capable of representing the physical nature of things, but not their spatiality or space itself. We can get an impression of the latter through binocular vision or the movement of the eyes, but this is an impression that assumes a curiously phantom-like aspect when acquired in isolation. This becomes apparent when watching three-dimensional projections. The spatial image is filled with more life when another aspect of vision comes into play, namely that of focusing on different distances. Thus, by means of our gaze, we can wander around virtually, in spatial depth, and only then do we realize that space is something in which we are.

This changes the scene. Space is genuinely experienced by being in it, through physical presence. Since the simplest and most compelling means of ascertaining our bodily presence in space is movement, those elements of vision that contain motion – changes of perspective and focal point – are best suited to conveying an impression of space. But seeing itself is not a sense that defines being-in-something but rather a sense that establishes difference and creates distance. There is another sense specifically for being-in-something; it is a sense that might be called "mood." A mood contributes to sensing where we are. By feeling our own presence, we feel the space in which we are present.

Our presence, where we are, can also be topologically understood as a determination of place. Indeed, sensing physical presence clearly involves both physical distance from things, whether they are oppressively close or very remote, and also spatial geometry, in the sense of a suggestion of movement, reaching upwards or bearing down. But a sense of "whereness" is actually much more integrating and specific, referring, as it does, to the character of the space in which we find ourselves. We sense what kind of space surrounds us. We sense its *atmosphere*.

That affects the perception of architecture. If it is true that architecture shapes space, then one must move about in these spaces in order to evaluate them. We must be physically present. Naturally, we will then look at the building and its construction, we will study its scale and shape, but such investigations do not actually require our physical presence. The decisive experience takes place only when we take part through our presence in the space formed or created by architecture. This participation is an affective tendency by which our mood is attuned to the nature of a space, to its atmosphere. And this demonstrates the truth of the proposition, ascribed to Polykleitos and explicitly noted by Vitruvius, that man is the measure of architecture – though in a different sense than originally intended.

Architecture and space

Peter Zumthor once said that there are two basic possibilities of spatial composition in architecture: the closed architectural body, which delimits, and the open body, which embraces an area that is connected with infinite space.[2] He apparently means such things as a hall, on the one hand, and a loggia or a square, on the other. But are those the only ways architecture composes space, or only the most fundamental? One need not even abandon the idea of space implied by Zumthor in this passage in order to imagine possibilities outside the dichotomy of delimiting and embracing. What does a medieval fortress do on top of a mountain? What did Jonathan Borofsky's *Man Walking to the Sky* do in front of the Fridericianum during *documenta 9* (Figure 13.1)? What does an airplane do in the sky? They concentrate space, open space, and create space. The fact that two of these examples are not architectural is irrelevant, since the possibilities they address apply no less to architecture. They merely stand on the edge of a geometrical or more generally speaking a mathematical treatment of space, in the transition to a space of physical presence. The classical, mathematically predicated concept of space dealt with two basic types: *topos* and *spatium*. Space as *topos* is a spatial locus, the space of contiguity and surroundings; space as *spatium* is the space of distance and scale. A medieval fortress on a mountain, for instance, creates a place; it articulates open expanses and concentrates space in one place. Here, one can see how the experience of this space is incorporated in physical presence. When we are in the vicinity, we sense that the space acquires orientation, a focal point, through the fortress. Borofsky's *Man Walking to the Sky* is simply an explicit rendition of what lines, beams, ledges, or ridge turrets do to space: they furnish it with a suggestion of movement, which is likewise not subsumed under the notions of delimiting or embracing. Suggestions of movement are inscribed in nothingness, as it were, and tend to open up space. Devices of

Figure 13.1 Jonathan Borofsky, *Man Walking to the Sky*, in front of the Fridericianum
during *documenta 9*, Kassel, 1992. Photo Gernot Böhme

this kind are not new to architecture. Think only of the flaming sweep of
Japanese roofs. An airplane in the sky also articulates space – creates space by
marking itself as a dot in the indeterminate expanse of the sky.

On studying these examples, one notices that they involve a concept of
space, or rather an experience of space that does not require things, whereas
the spaces of place and distance are essentially defined by things. But space as
the space of physical presence is at first nothing but a palpable, indeterminate
expanse out of which variously constituted spaces can be formed through
articulation. Orientation, suggestions of movement, markings are such forms
of articulation. They create concentrations, directions, configurations in space.
Since these articulations do not presuppose objective space but are rather
inscribed in a void, as it were, they must rely on the cognitive subject or, more
precisely, on the physical presence of people. It is the space of bodily feeling –
feeling that reaches out into indeterminate expanses – which acquires shape
through articulation of this kind.

Once it is determined that this space is fundamental to architecture, more
fundamental even than *topos* and *spatium* – because architecture does not
make buildings and constructions in isolation but for people – then it is easier
to accept architecture's involvement with non-classical, i.e. non-objective
means of constructing space, above all, light and sound. Light can create a
space, as in the cone of light from a street lamp into which one can step.
Sounds, noises, and music can also create spaces – self-contained, non-objective
ones – as is most impressively illustrated by listening to music with headphones. Architects have always made use of these means, but one has the

impression that the age of such devices has only just dawned. This may be related to the fact that by technical production of light and sound architecture no longer depends on the vagaries of seasons and days or festive occasions. Though Abbot Suger made architectural use of light as early as the twelfth century, such effects were, of course, dependent on the weather and the time of year. Today it is possible to make lighting and acoustics fixed constituents of architecture.

When one speaks of light and sound as aspects of spatial design, one thinks initially of their use as objective factors. In Axel Schultes's buildings, for instance, one finds capitals of light and walls of light. But this underestimates the spatial significance of light and sound, for they also create spaces of their own or give a space a distinctive character. Light that fills a room can make that room serene, exhilarating, gloomy, festive, or eerie. Music that fills a room can make it oppressive, exciting, or fragmented. The character of such spaces is experienced by the mood they convey, which takes us back to the beginning again, to atmosphere.

Conclusion: architecture or stage design?

Recognition of the space of physical presence as the actual subject matter of architecture moves perilously close to stage design. The latter has always known that the spaces it creates are spaces of atmosphere. And stage design has always made use not only of objects, walls, and solids, but also of light, sound, color, and a host of other conventional means: symbols, pictures, texts. All of these factors are relevant not because of their objective properties but because of what emanates from them, what they actively contribute to the scene as a whole and to the atmosphere with which it is suffused. Should the architect learn from the stage designer and perhaps even develop a new awareness of his art? Hasn't the architect always looked down upon the stage designer as a younger brother or even as a frivolous disciple of his own art? But why should this affinity between architecture and stage design be so threatening? Is it because architecture might melt entirely into the postmodern art of staging? That will not happen.

Life is serious; art serene. And that continues to distinguish architecture from stage design. Architecture does not build for the sake of the engaged or detached spectator watching a play, but rather for people who experience, in spaces, the seriousness of life.

Notes

1 G. W. F. Hegel, *Aesthetics: Lectures on Fine Art*, trans. T. M. Knox, Oxford, Clarendon Press, 1975, vol. 2, pp. 633–4.
2 Peter Zumthor, *Thinking Architecture*, trans. Maureen Oberli-Tumer, Basel, Birkhäuser Verlag, 1998, p. 21.

14 Staged materiality

Two bookshops

In the old city of Constance there are two bookshops, only a few hundred meters apart, but so different from each other that one could believe, going from one to the other, that one was entering another world or time. One, Das Bücherschiff, is approached by steps and a narrow entrance door. It gives the impression of being jammed into a residential house: one passes through rooms unclearly and crookedly connected to each other by steps. In actuality, the space is articulated not by walls, but by the wooden beams of half-timbering, whose spaces are filled not with bricks and mortar but with books. The beams are what shape the atmosphere of this bookshop: yellowish brown, rather soft wood, corresponding to the warm light of the incandescent lamps. The wood gives the effect of being well-worn, irregularly hewn, old, but not aged; rather, "matured." The atmosphere tempts one to linger, to rummage. One feels no sense of being observed; one could belong there oneself. In the summer, they say, one can also drink tea in the back courtyard of this bookshop.

As for the other bookshop, Gess: entering at ground level, one slides, so to speak, past the sale bins in front, through the glass door, and onto a "conveyor-belt" running at an angle through the entire store, in actuality a marble passage. Following it, one passes the stacks of sale books and best-sellers and the cash register, quickly finding one's way to the various departments clearly indicated on the walls. Toward the rear, the store widens onto a second floor, clearly marked, announcing its presence with a steel platform cutting into the marble path at an angle. Glass, marble, stainless steel, and metal surfaces clothed in elegant gray define the atmosphere here.

The visit, one feels, must be quick and decisive. Information is at stake, and this is the place to find it.

Are these different worlds? Yes and no. The spatial organization and above all the dissimilar materials do in fact produce a feeling of being in different worlds. One result of this difference might be that the two shops would attract quite different clientele and customer personality types. And for people of the sort who have "their" bookshop, the difference in the atmospheres of the two stores certainly will determine their priorities. But the fact that the two bookshops, as is usual, have different areas of specialization – the one perhaps more visual arts and literature, the other more design, languages, travel, and pop psychology – is basically irrelevant. The retail book trade as a whole is so outstandingly organized – there is probably nothing in the entire world to compare to the German book trade – that one can order practically any book in any store and receive it the next day. Functionally considered, all bookshops are alike: they are terminals of the major retail booksellers. But in their atmosphere they are not alike at all. On the contrary: their functional sameness permits and indeed necessitates the differences in their aesthetic presentation. Precisely because the differences between two bookshops can scarcely be articulated functionally, they must be articulated in the design. The competition is a competition of atmospheres: here wood, rusticality, and warm light, there chrome, glass, steel, and neon lighting. And so the initial assessment is reversed. For what appeared at first glance to be the old and the new, the conservative and the progressive, the difference between two worlds, proves upon closer examination to be the broad range of variation of a single world, the modern or post-modern. Both variations are the product of design, the conscious creation of atmosphere, a theater of, and by means of, materials.

Material and materiality

Considered more closely, neither in the one place nor in the other must the beams carry any load or the steel bridge provide access to something otherwise unreachable. Both the half-timbering and the steel are non-functional, i.e. purely aesthetic. Or, better said, their function consists in presenting themselves. They help shape the space – they represent themselves, half-timbering and bridge – or rather, their materiality, wood and steel. The materials are, so to speak, emancipated. Their functional liberation enables them to exit as pure appearance: they no longer have actually to perform what they promise, as long as the promise is there.

And so the old call for the integrity of materials is transformed into its opposite. Its demands had forced the materials into inconspicuousness: precisely because they were supposed to correspond to function – of equipment, of buildings – they disappeared behind it. The new sensibility for materiality prevalent in current design and architecture calls for exactly the opposite. Materiality is supposed to show itself, to come forward, to help shape the atmospheres in which we live. Material and materiality thus part ways, as do the processes of making and perception.

Material is the stuff of which things are made. Its qualities are inconspicuous; they don't call attention to themselves. What is significant about material is how it responds to manipulation and stress and, no less important, how it fits into legal and economic calculations.

Qualities that have to do with the manipulation of materials, for example, are malleability, ductility, fusibility. Qualities related to stress are elasticity, breaking strength, non-flammability. Qualities attributable to a material qua raw matter in economic and legal calculations are above all homogeneity, standardization, and consistency of quality. As such they allow for ordering according to list, price guarantee, and warranty. For the production of objects, this triad of working, stress-related and economic–legal qualities renders superfluous the question of what the objects are made of; the material, the raw matter, is defined not by its character, but by functional equivalents. This gives rise on the one hand to the dominance of "characterless" materials in current production – of particle board, concrete, plastics. It leads to the systematic construction of materials according to the qualities demanded. This is the birth of the science of materials as engineering technology: ceramics, alloys, crystalline structures, and sophisticated hybrids of all three are developed with great ingenuity for specific functions.

On the other hand, the character of materials becomes autonomous: materiality becomes pure outward form. Wood, glass, steel, and marble as elements of architecture and design no longer designate materials in themselves, but qualities of appearance, the more characteristic, the better. Wood may still be wood, but oak is certainly a veneer and red oak a stain. Decades ago, Jean Baudrillard spoke of the *valeurs d'ambiance*. Nowadays this phrase should probably be translated as "theatrical value."

Paradigmatic for the rift between material and materiality, between the quality of the raw stuff and its theatrical value, is particle board. But of course the discrepancy between surface and inner structure that it epitomizes has precursors reaching back far into the past: not only related veneer techniques, but also architectural facings, stuccoed marble, enameling. In fact, the materialist Semper's own opera house is a prime example of the split between material and the staging of materiality, or materiality as theater: the marble columns are stucco, the wooden paneling is painted.

This could bring us to the premature conclusion that the discrepancy between material and materiality is a perennial phenomenon, a part of culture per se, as it were. After all, weren't the Egyptians already masters of surface finishing, and wasn't the objective of 2,000 years of alchemy the semblance of matter, i.e. the production of materiality? There is certainly truth in this. But one has to recognize that the interest in materiality as the reality of appearance is tied to particular cultural and economically defined epochs; in short, to epochs of luxurious economy. And as far as alchemy is concerned: the absence of quantitative methods of description left no alternative but to define materials by the quality of their outward appearance. Only such a definition necessitates the warning that "not all that glitters is gold." Materiality is thus revealed as a product of economic development and of the state of science and technology.

The economy of developed industrial nations is dependent on the production of luxury articles. When basic needs are satisfied and production for war declines, the maintenance of production levels and, indeed, any growth at all, depends on the demand for luxuries and on their artificial – i.e. fashionable – or technological obsolescence. This leads to the dominance of the appearance value of products, of aesthetics over practicality. On the other hand, the development of science has deepened the rift between essence and appearance and has made the definition of materials independent of their outward form. In effect, the progress of technology has situated the level of human creativity ever deeper within the material. For the Greeks, the prototype of creativity was the craftsman giving a particular form to a given material (the carpenter, the stonemason). Today, the material itself is the actual object of creativity: what is created is its inner, not its outer form.

And so we have electronic devices in wood grain, marble-ised typewriters, express trains in white, gold, and silver, post offices in marble, department stores like castles. This development raises the question as to how materials are perceived at all, or rather – approaching it from the perspective of the object – how they present themselves. This could be related to the question of why, in the aesthetics of materials, traditional materials still dominate, i.e. why modern substances are attired in the design of traditional materials.

The manifestation of materiality

Materials manifest themselves with a particular character. What we designate as character is the structure of their atmospheric aura. That this character is decisive is shown by the fact that, for example, when one needs wood to create a particular atmosphere in a space – whether to achieve a sense of easiness and warmth or of prosperity and solidity – one has to give the surfaces wood-character. Under no circumstances does this simply mean that they must look like wood – although that too, to be sure. Here the texture comes into play, already giving us an important element of the character, that is, the specific way in which something displays itself. It may be the linear patterning, otherwise the grain or marbling – or to put it in general, if paradoxical, terms: the typical form of irregularity. The significance of this paradox is currently being studied in fractal geometry and chaos theory. The results are tentative and therefore uninteresting for the aesthetics of reception, but they could become important for the aesthetics of production, particularly when the design of types of irregularity is at issue. This gives us a preliminary answer to the question of the aesthetic prevalence of traditional materials. Up to this point, only the one or the other has been possible for human creativity: either the regular – from Plato's ideal bodies to the symmetry of wallpaper design – or the irregular – from the spontaneous idea to *peinture automatique* – with one exception: handwriting.

Handwriting – typical irregularity, non-conceptual recognizability. This is one way in which materials show themselves, one which nature performs for us in unending variety but which up to this point has scarcely been successful in industrial production, though perhaps it was never intended. But in order to let a particular materiality appear, it is not only important that the surfaces look like ... this or that. That would be colorless and flat, and would expose itself as an imitation. To be sure, whether it is an imitation or not is irrelevant when the appearance of materiality is at issue. But clearly there are further dimensions in which materiality characteristically manifests itself, without which an imitation would not be sufficiently watertight: for example, coloration in all its nuances, the microscopic structure of the surface, i.e. the degree to which it isn't surface at all, its haptic qualities, one could say. The crucial point is precisely that these qualities do not usually have to be verified haptically at all – they are atmospherically perceptible even without the concrete sense of touch. At a physical level this is doubtless connected to the optical features of the surface formation, to absorption, diffusion, refraction. But with respect to the aesthetics of reception, the issue here, as with color, is what one can call its synesthetic character: whether a material gives the effect of being warm, gentle, repellant, smooth, damp, obtrusive, or reserved. This kind of character always affects several senses, and for this reason can be perceived representatively through different senses, or, from the perspective of the object, can be produced through different material qualities: the cold through blue, the repellant through a glossy finish, the shrill through color contrasts.

If we call the first dimension in which materiality manifests itself its physiognomy, the second would be its synesthetic character. A third dimension is the social character. Materials have a social character to the extent that they, by reason of their use of culture or tradition, stand for something: the 1920s, for example, or coziness, grandeur, in some cases also the natural. But that brings us to a new subject.

The iconology of materials

Only recently has the discipline of art history become aware that such a thing as "The Language of Materials" exists. Inherently, the language of materials is surely as old as art itself. Up to this point, however, the study of art, with its emphasis on the language of forms, has not given sufficient weight to the fact that materials, too, are carriers of meaning. In any case, since iconology is the study of images, it would be better to speak of the semantics of materials. The semantics of materials is based partly in their origin, partly in the privileged access of particular strata of society to certain materials, but partly also in sheer convention, whether fashion or ideology. Of course it was significant that a material came from the Holy Land or that the stones of a particular building were fragments of the Bastille. And, for example, if the Roman emperor had a monopoly on Tyrian purple or a particular Egyptian porphyry quarry, then purple and porphyry stood for imperial grandeur. What is interesting is how ideology and convention can also invest a material with the semantics of origin, even when it is in fact found everywhere. In the nineteenth century, for example, granite acquired significance as the material of the fatherland, even though the German fatherland or Prussia is hardly distinguished by the presence of granite, but more by the absence of any other usable kind of stone. The language of materials in the history of art follows still other cultural patterns besides origin, privilege, and ideology: for example, the alchemistic classification of the seven metals or their hierarchical ascription to the ages of the world – the golden age, the silver, the iron. Through the use of a particular material, then, works of art can represent the world order or the hierarchical order of society implicitly, i.e. in addition to or through their sculptural form.

To be sure, the language of materials as identified by the history of art is only a very small part of the larger potential of materials for significance in the aesthetic constitution of everyday life. Here it is better to speak of a social character than of a language. From the perspective of art history, the effect of a material's social character presents itself as a code which must be deciphered. This is due precisely to the conventional nature of this character. The difference between a lion of bronze and one of granite is only perceptible to us at the synesthetic level. But when a viewer no longer shares the convention, the fact that the use of one material as opposed to another is symbolically something can only be mediated through art historical hermeneutics. In contrast, leather as a material for car upholstery exudes elegance, loden cloth conveys a sense of the rustic, while stainless steel is chic. These effects are directly perceptible, but their conventional character can fade and in some cases even turn them into their opposite. A typical example is the history of the aesthetic effect of concrete, "which in the first half of our Century was invested with positive, almost messianic significance, and in the meantime has degenerated into a popular metaphor for vices such as contempt for humanity, narrow-mindedness, and heartlessness."[1]

Sensing material

The creation of an atmosphere through the character of materials can indeed be called magic. What is magic? Conjuring, telekinesis, the triggering of effects through signs. Magic is puzzling, it is incomprehensible. Because cause and effect are not of the same kind, it is dangerous and insidious: it can also work against our will. All of this applies to the effect of materials in the theater of the world in which we live. Most remarkable and incomprehensible of all is how this effect can be had through mere appearance, i.e. through materiality without material. In fact, the pure aesthetic of materials assumes we won't handle or touch them. What produces an atmosphere of coldness or softness would probably be robbed of its effect if one tried to verify its promises by touching. On the other hand, it is at a very physical level that the synesthetic character, or the character of the atmosphere produced, moves us. Even simply to confirm the qualities of a material on its surface would cause the atmospheric effect, for which it is in fact employed in design and architecture, to collapse. The effect is deep and subcutaneous, as a rule even unconscious. Only afterwards, when we already feel a certain way in a space, when the atmospheric effect of the materials has already completely bewitched us, do we perhaps try, irritated, to identify its origin. This is what is eerie and dangerous. The same is true for the social character of materials. As opposed to the discipline of art history, we have asserted that this character is not usually read, but is sensed. The noble, majestic quality of a material, its elegance or old-fashionedness are sensed. But this does not mean merely that the material is able to point to or signal the noble, the majestic, the elegant, or the old-fashioned; rather, it seems to radiate them. They must in some way be connected to, anchored in, its material qualities. This is why it is sometimes difficult to distinguish clearly between the synesthetic and social character of a material. Is solidity synesthetic or social? Probably both. The solidity of the material stands for the solidity of the social conditions in which it is employed.

Materiality can certainly be used to make magic. Designers, interior decorators, set designers do it. But what are they relying on? What gives them the certainty that their magic will work and that their conjurings will reach the public? How is it possible for us to be affected physically by something with which we have no physical contact at all? Magic?

The answer to all these questions probably lies in the fact that, besides the perceptual and work-related relationships, there is a third relationship to material which we will call the medial. In the working relationship we are involved with the material as raw matter. When we grapple with it, intend something with it, seek to form and change it, certain qualities become manifest: material is elastic, soluble, fusible, and brittle. In the perceptual relationship we are involved, not with material qua working stuff, but with materiality, the pure form of its appearance. Here we encounter its physiognomy, its synesthetic and social character. But we can also be inside the material, walk on it, sit on it, rest in it and – eat it.

This relationship to material is dominant in early childhood, before the working relationship and distanced perception have developed. The fact that we exist as bodies among other bodies and live physically within different media is the basis of our direct physical experience of materials. We experience softness or hardness, wetness, dryness, coolness, and warmth on, or better, in our own bodies. Aristotle designated this special perception as the actual touching (*Haphe*). A more precise translation would be sensing. To sense a material is not to take note of its qualities by touching its exterior. This would involve more than just the surface, as Aristotle says, our flesh is simultaneously the medium and the organ of this sensing. Thus we experience and recognize firmness, softness, warmth, and coolness in our experience of ourselves. The sensing of materials is in this way a sensing of oneself. In this physical sensing of ourselves lies the foundation of the later perception of materials as well. To a certain degree it is never lost, for we remain bodies among other bodies and live within media. But the more distanced approach to material and materiality preserves these experiences only as a background memory. The magic of the material is disclosed, even if we inevitably fall for it again and again. And why not? How impoverished life would be without this element of everyday regression.

Note

1 Thomas Raff, *Die Sprache der Materialien. Anleitung zu einer Ikonologie der Werkstoffe*, München, Deutscher Kunstverlag, 1994, p. 15.

15 Architecture: a visual art?[1]

On the relationship between modern architecture and photography

A natural alliance?

The very first photograph was an architectural photo.[2] Since long exposure times were needed, architecture was an ideal first choice owing to its static nature. You ask yourself whether this served to forge a common bond between photography and architecture. Whatever it depicts, a photograph freezes its subject – and in what instances would this be more appropriate but in the case of architecture, which stands unshakably there. Indeed, the essence of architectural photography was often seen as presenting the subject as a timeless object, and of course this also meant to eliminate, if possible, and touch up, if need be, the insignia of the ephemeral that are found on and around buildings. What is remarkable is that within a relatively short period, these photographs assume the aura of the past, in other words, architectural photos relatively quickly become documents of architecture history. The most famous examples are the photographs of industrial plants by Bernd and Hilla Becher. They are not photographs of what now is but of what once was, documents of industrial archaeology. And this was arguably the intention, after all: the Bechers knew they were documenting something in the process of disappearing, namely the legacy of the coal and steel industry.

What is remarkable is that many photographs of modern architecture, in other words, of buildings that cannot be said to be technically obsolete, likewise appear to be documents of the past. Yet you could impute a totally different bond between architecture and photography in the modern age. For a period that was relatively brief, we see from our current perspective that architecture and photography come together in the aura of modernity. While the Bauhaus had arguably at the start of modernist architecture for the first time abandoned everything connected with the past, to rely on clear lines, functionality, and rational–industrial buildings, these concerns converged in photography, which was liberating itself from its dependence on other arts, above all, painting: no more pictorialism, no genre scenes, no staffage – the new goal was to show clearly and patently what photography was. Today it is difficult to separate the architectural developments of the 1920s and 1930s from their depictions in the black-and-white photography of the time. It would seem that the photographers and architects essentially wanted the same thing. Simple shapes, clear views, rationality, functionality. Art and design historian Gerda Breuer would argue that this appearance is deceptive. She ascertains that Gropius focused on "representations of external architectural elements" and photographs aimed at asserting architecture for propagandist reasons.[3] Breuer expounds on the difficulty Lucia Moholy experienced with her contract to photograph the Dessau Bauhaus in a manner that demonstrated its architectural structure. After all, as Breuer wrote:

> One fundamental aspect of Modernist architecture was that the building has precedence over the image, that it develops from the inside out, from the ground plan and functions and not from the elevation; its design is not conceived substantially from its external appearance. It follows that the core has precedence over the shell, indeed, the image of the building is essentially a derivative of its constitutive architectural properties.[4]

These observations are so sobering that we are obliged to cast around for another reason for the close relationship between modern architecture and photography.

To my mind, the reason lies in the development of architecture as a profession. Accordingly, the origins of the relationship between architecture and photography is not so much to be sought in the items as in the architects – and it is only as a consequence of this that the products of modern architecture have an intrinsic relationship to photography. The profession of the modern architect, who is no longer attached to an institution but operates as an entrepreneur, requires that he is identified through his works. However, if he no longer only designs regionally – and today all the great architects are global operators – this can no longer be achieved through the works themselves but only through their photographic representation. The architect builds his reputation through the photographs that are taken of his works and their dissemination in illustrated books, architecture magazines, and the art sections of newspapers. In her contribution to the catalogue for the ninth International Architecture Biennial in Vienna, Nanni Baltzer pointed out that this is why many architects have entered into firm cooperations with renowned photographers. It is in this cooperation that their work finds its perfection. She cites Jacques Herzog's remark that "tout, au monde, existe pour aboutir au livre."[5] Renowned architectural photographer Julius Shulman has even claimed that the final product of architecture is the photograph of it: he argued that "once the final product had been attained and the building was 'ready to be photographed' the time had come to present it to the media. Its historical value was then appraised alongside its market value."[6] However, the architect's obligation to use photographs to present his work not only relates to the finished building, it creates a structure for all the preliminary stages from design and acquisition through to development. However, today, apart from the drawing and of course the computer simulation, the model also plays a role. But despite having a closer relationship to the finished building, such three-dimensional models by their nature serve a purpose similar to photographs in that they provide a visual representation of the planned and later of the finished building.

It would seem that we have touched upon the core of the matter here. Since architecture is a visual art, its depiction in visual media would seem to be an appropriate means of public presentation. But precisely this theory needs to be questioned: is it the central task of architecture to produce elevations? Only when doubt is cast on this does architectural photography truly become interesting, namely, when it faces a challenge, when it must portray something in architecture that is not immediately visible.

The fact that some photographers have recognized this task is evidenced by the aforementioned essays by Gerda Breuer and Nanni Baltzer. But the majority of architectural photographers follow the classic maxims as they can be found say in Shulman's work. Specifically, the product of architecture is to be seen as something like a large sculpture, which is to emerge in form and contours as clearly as possible in the photograph. It follows that it must be photographed under conditions that are as ideal as possible.[7] The idea is to have the overall design emerge more clearly,[8] and consequently you need to select the most favorable weather conditions and incidence of light that clearly articulates the contours of a building, that if necessary you employ artificial lighting in order to avoid – apart from the classic central perspective – everything that might be reflected in the image as a subjective treatment. In particular, the horizontal lines must be retained and there should be no veering lines.[9] Since this type of photography ultimately aims at representing the building as such, the photographer seeks an unrestricted view of the object or creates it or even touches up distracting staffage at the end. This also includes the people, in other words visitors to or residents of a building. In these "normal" architectural photos they come across as accessories or props and scarcely differ from the tiny figures the architect uses to decorate his models with along with small trees and imitation grass. Since the gaze of the observer in the photograph is directed toward the architecture as subject, the people appear like display dummies in store windows.

Criticism

Every two years there is a European competition for architectural photography. Since the photographs submitted for this competition need not act as marketing for a specific office, you might hope that among the winners you would find another type of architectural photography that differs from the usual standard, especially as participants are given a theme, which serves to locate architecture in other contexts and perspectives. But that is not the case. Even in 1995, when the topic was *Mensch und Architektur* (Man and Architecture), the first prize went to Petra Steiner and Grit Dörre for photographs of Portakabin settlements. However, they by no means showed life in or around the Portakabins, but simply presented a heavyset female resident standing with arms folded in front of it, essentially in portrait format. The image does not contain a relationship to the buildings or the surroundings – not even a negative one. Others in the same competition portrayed people making orchestrated gestures, such as pointing out certain defining characteristics of the architecture or, in one example, showing neighbors on their balconies handing each other bottles of beer thanks to the close proximity of the buildings.[10] Nothing ephemeral can be seen on the images, insignia that point to the time of day or season are lacking.

In the accompanying text of the respective publication, author Heinrich Klotz says of the photographs (which he found in a book on European architecture) that they were at pains "to achieve an unrestricted scenery, in order to show the building to full effect with the minimum of distraction. The interiors are swept clean and the squares in front of the facades untouched by human foot. Depicted are buildings such as you have never seen as if they were behind glass."[11] Evidently, the criticism in this comment led to the competition *Mensch und Architektur* – though it did not succeed in changing the situation fundamentally.

The decisive point is that the relationship between man and architecture is not manifested by placing people next to or in the buildings, but rather by introducing to the image the experience that you have as a person in the vicinity of or inside architectural structures, whether through figures that allow you to experience the scene or through the perspective, which makes this experience palpable. However, this presumes a basic criticism of architecture as a visual art. Such criticism is made by Finnish architect and architectural theorist Juhani Pallasmaa in his essay *The Eyes and the Skin: Architecture and the Senses*[12] in which he discusses ocularcentrism in Europe. He argues that not only the depiction of architecture, but also the architecture itself has fallen prey to this dominance of vision, and attempts as a countermove to demonstrate the importance of the other senses and ultimately makes a case for the physical and forms of life. With buildings such as Alvar Aalto's Villa Mairea he attempts to illustrate a turn in architecture. But: "until recently, architectural theory and criticism have been almost exclusively engaged with the mechanism of vision and visual expression."[13] Alongside the characteristic European cultural dominance of vision, he also sees the intensification of this dominance, which was achieved during our historical period through the dependence on technological developments and market events:

Art and Architecture have adapted the psychological strategy of instant persuasion familiar from advertising, and buildings have turned into image products detached from existential sincerity. As a consequence, architecture is turning into the retinal art of the eye. But the change goes beyond mere visual dominance; instead of being a situational bodily encounter, architecture has become an art of printed image, fixed by the hurried eye or the camera.[14]

In this comment Pallasmaa has simultaneously formulated the crucial aspect in experiencing architecture: "bodily encounter" not, however, how it can be captured using the medium of photography. If we look at the presentation in book form of the paradigm he cites for an alternative architecture, namely that on Villa Mairea, some of this paradigm's requirements are already met: the building is shown here in various seasons, there is a particular focus on the materials and details, the interplay of artificial and natural light in the image is not blurred – in other words, the times of day remain visible. Moreover, the magnificent color photos especially produced for the publication are juxtaposed with historical black-and-white images, which at least in part visualize the life of the original developers and owners. A new departure in this context is a photograph in which together with Maire Gillichsen and Aino Aalto, who sit and stand near the opened sliding wall of the living room, the viewer's gaze is drawn out onto the garden and park. This kind of composition, arguably introduced in Caspar David Friedrich's *Mönch am Meer* (Monk by the Sea), takes the observer into the view that opens up, and consequently allows him to participate in the experiences of the persons present.[15]

It is characteristic that the critics of traditional architectural photography tend to look for alternatives in another class, namely in artistic photography – in other words, in those photographs in which although architecture may be the subject, the images do not serve a representative purpose or act as advertising as architectural photography generally does. Gerda Breuer refers to a series of photographs that Thomas Ruff submitted for a catalogue on the period Mies van der Rohe spent in Berlin,[16] but also to photographs of buildings by Swiss architects Herzog & de Meuron that were featured at the 1991 Vienna Biennial. Ruff's photographs, or at least these particular ones, are characterized by a lack of sharpness that encourages an atmospheric perception, but also by the fact that they were digitally manipulated. Ruff emphasizes that his intention is not to depict an object: "My prime concern is not a phenomenology of the photographed item but rather producing an image."[17] His comment might make you doubt whether this method can act as a model for an alternative architectural photography, since after all it was to be more phenomenological than traditional architectural photography. The latter aimed to portray the architect's visual idea, which means eliminating phenomena that are ephemeral in perceptual terms. This explains why classic architectural photos frequently appear Surrealist and it is more than logical that photographer Gerald Zugmann produced his photographs using the architect's model rather than the finished building.[18] It is presumably more the out-of-focus quality, the blurring in some of Ruff's photos, that makes them models for a new kind of architectural photography. In the aforementioned essay, Nanni Baltzer wrote of "atmospheric architectural photography" and consequently a return of pictorialism and a new romance in photography. She not only finds these qualities in Ruff but also in Andreas Gursky,[19] Luisa Lambri, and, naturally, Hiroshi Sugimoto. In a series of permanently exposed architectural photos Sugimoto produced what was in retrospect arguably the most striking image of the Twin Towers in

Manhattan. They appear on his photograph like shadowy silhouettes, threatening in their ghostly yet powerful presence. According to Nanni Baltzer, this return of aura is achieved by a blurring technique or lack of precision. In this she cited a remark by Herzog saying that the indefinite and vague in architecture makes it much easier to communicate the architect's idea than a detailed model: "Herzog's comment on the immediate connection between imperfection and aura also applies to photography."[20] However, this cannot be the whole truth.

Space, the scenic, and photography

If we assume that it is the task of architecture to shape or even create space and more specifically space for people, this eliminates the natural assumption that photography is the ideal medium for conveying the essence of a specific example of architecture. After all, you cannot photograph space as such. This not only derives from the fact that photographs are two-dimensional; in addition, what is depicted on a photograph, the things, the buildings, are only the delineations of space or markings in space but not the space itself. Moreover, what you can at best convey in the photo using the theories relating to portrayal is space as a medium of depiction, while what really counts for the evaluation of architecture is space as a space of physical presence.[21] The best method of experiencing architecture and its value is to move close to it or, even better, to enter the building to find out what it feels like from the inside. Defined space – the aura of a building – can only be conveyed indirectly in a photograph, through the elements that are relevant for the production of its atmosphere. However, the statement a good photograph makes about architecture would have to be an indirect one; the photograph should not primarily be a portrayal of the building but rather make it possible for you to experience things through the photograph that you should really only experience in person, near to or inside the building. We have arrived at the maxims for good architectural photography. We can express the same with a comparable remark Umberto Eco made when referring to the task of a drawing: a photograph should not be a depiction of reality but should convey experiences of reality using symbols. In this context, "symbols" refers to those elements either in reality or in the image that are responsible for creating a building's characteristic atmosphere.

Bearing this maxim in mind, it is hardly surprising that the critics of classic architectural photography should select as an alternative those examples of artistic photography whose primary intention was not to portray a building.

What makes architectural photography an art is the fact that it sets out for itself the task of making visible something that cannot be visually perceived. The latter is crucial to architecture in its space, the space of physical presence. Space is not an object but rather the background, the horizon, room to move, in which objects can appear, it is actually the void – and something crucial to experiencing it: it is everything. Being in space means being surrounded by space. But this is the experience that is not given in a photograph, unless it is indirect – as if the observer were taken like an avatar into the space of depiction, namely, the image. This is precisely the effect Heinrich von Kleist mentions in his review on Caspar David Friedrich's *The Monk by the Sea* (1808–10). Just how difficult it is to take the viewer into the image is manifested in Gursky's photographic repetition, if it were conceived as such – of Friedrich's Monk: the figure in the image is too much a figure that you observe and not one through which you observe the Monk.

It ought to be easier to bring the spatial suggestions of architectural elements into the image – for instance, by using cantilever girders, which open up the breadth of the space, or encircling elements, which mark the divide between inside and outside. The latter will be all the more successful where the enclosure is not complete but where through the gesture of confinement the outside becomes visible. Similarly, situations involving movement, which in reality are given through paths and routes – for example, through the curve of a staircase or a house gable – can be traced in the image via the corresponding lines, because in the image the eye wanders along them as in reality. We cannot imagine a line differently, says Kant, than as we draw it in our minds. In order to replicate the characteristic play of movement in a building, which in reality you can sometimes only compose into a whole by moving yourself, it might occasionally be a good idea to create an architectural photo as a photomontage. In his own architectural photos Gursky further refines this method using digital manipulation. But the technique occurs as early as the 1920s, for example in the depiction of a Cologne office building.[22] There is also an affinity to Pablo Picasso's portraits in which he unites views of a person from various perspectives in a single image.

The phenomenon of spatial thrust is difficult to convey using a photograph because the photo is conditioned by technical constraints that run counter to three-dimensional thrust. Of course, one method of introducing spatial thrust into an image would be to have it coincide in the photo with the direction of the gaze in terms of perspective. But the confusion of competing directions, which is the experience of the visitor, is abandoned in favor of a single direction, but this method can also be dysfunctional. For instance, if you wish to reproduce the feeling in Gothic cathedrals of being pulled both upward and outward by directing the camera toward the ceiling, the narrowing perspective will fail to produce this experience.

Naturally, the manner in which a building is experienced is also largely determined by the views it offers. In this context classic architectural photography, which basically treats buildings like large sculptures, comes into its own. Moreover, through photographs, in other words optically, you can reproduce very well the function a building may fulfill as a concentration of its environment. However, for the experience of the visitor or resident of a building, the perspectives that a building enable are equally important. In this sense it is to be welcomed that a prize went to a series of photographs in European architectural photography, which gave the views looking out windows from various floors of a building – and they were photographed from the interior.[23] The entrances and exits of a building – the openings – but also the views from beneath arcades or protruding roofs, are essential for how space is experienced, because in each case they set the presence in a place in relation to the surrounding space. This makes it particularly clear that what is often decisive for how a building is experienced is not only the view of the building but also its embedding in the setting. A characteristic example of this is the Kultur und Kongresszentrum Luzern (Lucerne Culture and Congress Center), which is in fact only a simple glass cube, but thanks to the overhanging roof is placed in relation to the lake and the high mountains along its banks.

You might summarize what has been said so far under the requirement that architectural photography should demonstrate in what manner a building creates a location in space or opens up space at a specific place. However, this still falls short of what might be summarized under the caption of "atmospheric photography" as defined by Nanni Baltzer. Above all, it lacks the temporal aspect, the ephemeral. We are astounded to read that Walter Gropius attached great importance to his photographs showing the building to be "currently existing." However, in actual fact these modern (for their time) photographs of modern architecture are more suited to presenting their objects as eternally existent, something that could also have been achieved using models. There is hardly any evidence of historic time. The weather is the eternally favorable weather offering ideal lighting, nothing indicative of season is visible, and there are no chance scenes. But it is such chance aspects that lend the permanent building the flair of actually existing. This opens the doors to the arbitrary, the random, and you wonder whether in this manner those elements that constitute a building can be better communicated to a person who never had the chance to experience it himself through actual physical presence. Would not this incorporation of the ephemeral result in favoring the photographer's subjective view over a more objective portrayal? An alternative would be to create a series of the respective building based variously on season, time of day and weather conditions in the way that the artist Claude Monet produced four paintings of Rouen Cathedral that showed the church in the changing lighting of a single day. We have a similar example in the aforementioned book about Villa Mairea showing the building in various seasons. But even without this, contrary to the criticism by Heinrich Klotz cited above, the architectural photos that incorporate the ephemeral would depict the buildings as they had at least been seen once or can occasionally be seen. It would then be possible for the viewer looking at a photograph, which could also be called a snapshot of the building to add in his imagination the temporal aspect. As is generally known, this is best achieved when a photograph captures a "decisive moment."[24] In the debate over the concept of *ut pictura poiesis*, Lessing had already pointed out in his treatise on Laocoon works of fine art are better suited to depicting physical conditions, and could integrate the temporal precisely in those situations when they portray an event that is about to cease. The same applies to architectural photos. They are best able to communicate the scenic experience a visitor to a building has when they show the building itself as a scene in which life occurs. You could say that the photographer should see a building or an ensemble of buildings through the eyes of a set designer. Then all of the following would play a role: if you were to consider the incidence of light, the color of light, season and time of day, and the materials in their synesthetic qualities, you would ask yourself which characteristic scenes of life a building actually permits. In this case, the aim would no longer be to present a building in its precise contour but rather to convey an impression of its weighty existence. It will not be important to wait for ideal lighting conditions or produce artificial light, but rather to show what a building looks like in the play of chance light. Finally, it will not be important to see the building as precisely and complete as possible but rather to make it perceptible as the basis for the spatial experience one has either in it or when in its vicinity. Ultimately, a building will not be created by the architect for the photograph but rather as an element in an environment in which it allows residence. Capturing the mood in a photograph means communicating the atmosphere an architect generates through his building.

Notes

1 First published in Ralf Beil, Sonja Feßel (Hg.), and Andreas Gursky, *Architektur*, Ostfildern, Hatje Cantz, 2008, pp. 24–31.
2 Michel Frizot, ed., *Neue Geschichte der Fotografie*, Cologne, Könemann, 1998, p. 21.
3 Gerda Breuer, "Architektur und angewandte Fotografie. Vom Bildgebrauch der Moderne," in Ingeborg Flagge (ed.), *Jahrbuch Licht und Architektur 2003. Architektur und Wahrnehmung*, Darmstadt, Verlagsgesellschaft Müller, 2003, p. 103.
4 Ibid., p. 101.
5 Nanni Baltzer, "'The Cloudy Translucence Like that of Jade ...': Atmospheric Architectural Photography," in Nanni Baltzer and Kurt W. Forster (eds.), *Metamorph. 9. International Architecture Exhibition*, exh. cat. *La Biennale di Venezia* (New York, 2004), pp. 94–109.
6 Julius Shulman, *Architektur und Fotografie*, Cologne, Taschen, 1998, p. 16.
7 Ibid.
8 Ibid., p. 286.
9 Shulman achieves this by aligning the rear plate of his special camera with the main plane of the building.
10 Wilfried Dechau (ed.), *Mensch und Architektur. Europäischer Architekturfotografie-Preis 1995*, Stuttgart, DVA, 1995.
11 Heinrich Klotz, "Über das Abbilden von Architektur," ibid., p. 34.
12 Juhani Pallasmaa, in Ingeborg Flagge (ed.), *Jahrbuch Licht und Architektur 2003*, Darmstadt, Verlagsgesellschaft Müller, 2003, pp. 62–87.
13 Ibid., p. 68.
14 Ibid., p. 69f.
15 Juhani Pallasmaa (ed.), *Alvar Aalto. Villa Mairea*, Helsinki, 1998, ill. 198, p. 88.
16 Terence Riley and Barry Bergdoll (eds.), *Ludwig Mies van der Rohe. Die Berliner Jahre 1907–1938*, exh. cat. Museum of Modern Art, New York, 2001.
17 Cited by Breuer, in Flagge, *Jahrbuch*, p. 100.
18 Peter Noever (ed.), *Gerald Zugmann. Blue Universe. Transforming Models into Pictures. Architectural Projects by COOP HIMMELB(L)AU*, Ostfildern-Ruit, Hatje Cantz, 2002.

19 This forges a link between Andreas Gursky's *Düsseldorf Airport II*, 1994, and Caspar David Friedrich's *Monk by the Sea* (1808–10).

20 Baltzer and Forster, *La Biennale*, 2004, p. 99.

21 Gernot Böhme, "Der Raum leiblicher Anwesenheit und der Raum als Medium von Darstellung," in Sybille Krämer (ed.), *Performativität und Medialität*, Munich, Fink, 2004, pp. 129–40.

22 Office building of the Hotel-Disch AG. Photomontage made by Theresa Schmölz, featuring individual photos taken by her husband, Hugo Schmölz, in 1929. Karl-Hugo Schmölz and Rolf Sachsse (ed.),*Hugo Schmölz. Fotografierte Architektur 1924–1937*, Munich, Mahnert-Lueg, 1982, Fig. 11.

23 Stefan Freund received a prize in the 1999 European architectural photography competition for his series of building floors. See *Deutsche Bauzeitung 4*, 1999, special issue, "Architecture in Context."

24 "Painting, in her co-existing compositions, can use only one single moment of the action, and must therefore choose the most pregnant, from which what precedes and follows will be most easily apprehended." Gotthold Ephraim Lessing, "Laocoön," 1766, in *Laocoön, Nathan the Wise, Minna fon Barnhelm*, New York, William A. Steele, 1970, p. 55.

16 Metaphors in architecture – a metaphor?

Architecture as language

It was the architectural theorist Charles Jencks, in his writing about postmodern architecture, who underscored the importance of metaphors for architecture. While recognizing in the architecture of his time a lack of the acknowledgment of the importance of metaphor, he thought that that would change, because "metaphor plays a predominant role in the public's acceptance or rejection of buildings."[1] Now it can be said, that Jencks himself was caught up in the fashionable theoretical trends of the time, looking at everything through the lenses of semiotics. His proclamation of a new epoch for architecture, that of postmodernism, has to be seen in relation to its understanding of architecture as language. The understanding of architecture through the metaphors of another art/discipline – in this case literature – arises because of the quite strange, but at the same time classic embarrassment to state what architecture should be as discipline in its own terms.[2] A discussion about a work of architecture is often conducted through references, comparing it to a sculpture, a painting, a musical composition, for example a fugue, or a poem. This should be questioned, given the common knowledge that architecture is mainly a spatial art. But one hesitates to state this quite simple truth and to use it in the description because space alone does not communicate what one associates with architectural space that is the aesthetic qualities and the emotional emanation of its works. The reason lies in the fact that our conception of space is strongly influenced by geometry, by space *formale Anschauung*, as Kant would say – i.e. space as formal intuition or as a medium of three-dimensional representation. While the architect indeed has a lot to do with this geometrical space – he has to arrange his works in the context of things and therefore to consider distances and volume, and even more, he has to use space as medium of representation for his designs, such as through drawings, models, and simulations. Yet, what matters in the end is the space in which we live, the bodily space. Each work of architecture creates or constructs a space, in which we, the visitors, move and in which we feel something. Thus, the architect works with the geometrical space, but he uses it to design the space of our bodily presence,[3] he determines the premises of bodily space experience, in short, of our feeling inside the space. If one makes reference to this space, the space of bodily presence, the experience and the effect on visitors is always already implicit and there is no need for metaphors.

Although the discussion about metaphors in architecture stems from the – remediable – embarrassment to talk in general about architecture in its "own" terms, it is necessary to question whether one can speak of a language of architecture, of architecture as language.

What are metaphors?

The term *metaphor* stems from linguistic theory, to be more precise, from the theory of literature. It was introduced by Aristotle in his *Poetics*. This occurs in a paragraph were he discusses in particular words and their use. The metaphor is not a class of words but rather a particular usage of words.

Its definition is: "A metaphor is the application of a noun which properly applies to something else. The transfer may be from the genus to the species, from the species to the genus or according the rules of analogy."[4]

The types of transfer are not interesting for us in this context; Aristotle introduces those types considering the question of how the transfer logically is arranged. It should be only mentioned that the modern use of metaphors is not reduced to the last type, which is analogy.[5]

What is crucial for us is that the transfer is always transferring to another object or discipline than the one which the word in question is originally related to. A classical example, used by Aristotle in his *Rhetoric*,[6] comes from the *Iliad*:[7] there, Homer names Achilles, the way he attacks Aeneas, a lion. Here the term *lion* is transferred to a man.

It is very important to underline that the transfer implies not only a dislocation on the level of the signifier, that Homer instead of saying *Man*, uses the term *Lion*. That would be simply false – because not a lion penetrates in Aeneas, but the man in Achilles. Homer on the contrary wants to say something extraordinary about Achilles, by calling him a lion or better, to let him appear in a particular light. He wants to articulate his audacity, his recklessness.[8] Thus we arrive at the question of the cognitive function of metaphors.[9] It is namely insufficient to consider metaphors only as an adornment of speech, as Aristotle does, because he treats them mainly inside his *Rhetoric*.

The application of metaphors allows seeing the object, on which the word in question is transferred, in a particular light. They articulate aspects of this object that in simple denomination would not be cognizable. We can even demonstrate – to stick to the example: "Achilles the lion" – that the Greeks only became aware of the character and mood of somebody through analogies with animals.[10] Therefore: courage is what one sees in the lion, timorousness in the deer.

The cognitive function of metaphors would not be understandable or would even be unnecessary if objects were already given to us concisely articulated as to their properties and structures. This is not the case. Rather objects are normally opaque, or one could say, they are given in a compact way.

The metaphor, i.e. the word, which is transferred on this object from another field of reference, transports a scheme, which organizes the perception.[11] In the example: the movements and gestures of Achilles are brought to the focus by the scheme *Lion*. Thus the metaphor does not apply a signifier to an already completely definite object but rather the representation of the object is organized through the metaphor. The proximity of metaphors to models in science stems from here: they allow for an initially diffuse amount of data to be theorized. Therefore, it is clear that the term *metaphor* – through the 2,000-year-long history of the theory of metaphors – has experienced an expansion of its domain.

But it has to be said: metaphors always have their place on the level of the signifiers, that is, of language, but these signifiers are metaphors only in a particular relation to the signified, to the things. We can therefore hold onto the following: metaphors are a phenomenon of language that however only appears when a discourse is a discourse about something – or toward something or somebody, to include also the convivial address and the insult. This will become important for the discussion of metaphors in architecture: if architecture is a language, what then is it talking about?

Metaphors in architecture

If one wants to judge, what role metaphors play in architecture, so one has to decide, if the focus is on the discourse or on concrete works of architecture.

Metaphors in the discourse on architecture

If one speaks of metaphors in the discourse on architecture, then one doesn't need the large-scale hypothesis of a Jencks, that architecture is a kind of language. The metaphoricity, then, consists only in the use of concepts in the description of works of architecture that stem from other domains. Such linguistic behavior can stem from the above mentioned embarrassment of not having a true conception of architecture as a particular art but it can also help to render more articulate the effect a work of architecture provokes. Obviously many examples, introduced by Jencks as architectural metaphors are, to be precise, metaphors in the discourse about architecture. If one calls the concrete grid of a multistore parking garage a cheesegrater,[12] this is a purely linguistic procedure: were one explains to somebody else short and concisely the impression of this façade.

In fact, this perspective will articulate the representation the parking-silo and at the same time imply a negative connotation. Equally there is a way of calling Jörn Utzon's Opera house in Sidney as "turtles making love," which is a purely linguistic metaphor. It manages to clearly explain to somebody the uncommon shape of the roof and at the same time to express the surprise of the observer in front of this work. But to call it a "mixed metaphor" as Jencks does, because it symbolizes "the growth of a flower over time – the unfolding of petals, fish swallowing each other"[13] is questionable. This argument seems to stem rather from the need of the architectural critique to look everywhere for meaning. In any case, it presumes that architecture is a language. Should we assume that the architect Jörn Utzon wanted to communicate something to the later observers of his works? This hypothesis would be dismissed by the ambiguity of the alleged symbol. Utzon has rather created a spatial form and therewith shaped the space of the Opera house and its context. Through this, the observer feels in a particular way, he experiences certain impressions. We could say that the shape of the roof creates a certain atmosphere.

Atmospheres are spaces with a certain mood.[14] In order to characterize the atmosphere of a building, it seems that one needs again certain metaphors. So for example one could say to define the suggestions for movement contained by an architectural form, that they are *emergent* or *rapturous*, or even *sublime*. Yet, it is important to hold onto the fact, that these are not metaphors – that would imply that the corresponding terms would come from somewhere else. This is not the case. Rather the characters of atmospheres are attributed to the atmospheres themselves and to the objects that produce them. In fact, they predicate properties of things not as their determination, but as their ecstasy, as well as the impressions, that a visitor will experience. One can call a man stormy, or the weather or the shape of a roof, in any case it concerns the impression that people experience in their proximity. Particularly revealing are the examples of *bright* and *sweet*. To call a valley or garden scenery bright is not the transfer of a condition of the mind to an object of nature – that would be absurd: should a valley have a soul? – but the characterization of one's own condition while looking at the valley. In the case of sweet one can even demonstrate that the term – at least in German – originated not in sugar-sweated aliments, but in the pleasant par excellence.

Therefore, there exist a lot of terms for the characterization of works of architecture, that appear to be metaphors, while in reality, they characterize the humors and dispositions, the synesthesias that one experiences in the environment of these works. However, symbols also belong to the creators of atmospheres, but not symbols in the way Jencks has them, as he equates them with metaphors. Symbols are conventional signs. Those signs thus have to be historical and socially habitualized. There exist many elements of Christian architecture, for example the cross or the gothic ogive, that applied somewhere else, imply a religious atmosphere. Therewith we have entered the discussion on whether there is in architecture that is in its works a use of forms that we can reasonably call metaphorical.

Metaphors in architecture seen as language

Without any doubt there are also processes of communication in architecture, and it is not mistaken to say that architecture turns toward a public through its work. Therefore, the diagnosis of Jencks, that postmodern architecture – in contrast to modern architecture – is more turned toward the general public. But not every communication is mediated through symbols – one has only to think about bodily communication the way the philosopher Hermann Schmitz discussed it in his phenomenology. Linguistic communication though is communicated symbolically – precisely by meaningful word – and normally it is related to an object. I say: normally, because exclamations don't need an object. Direct addresses namely don't have an object, they have an addressee, thus again we have here in the metaphorical speech the transfer to something that is outside language. Still, statements need a statement-object. But the object of a statement doesn't need to be a sensual object; it can also be an idea, or any other abstract. It is probably something like this that Jencks has in mind, when he considers architecture as a language: namely a communication mediated by symbols. Without doubt, we have examples for that. But it would be a pity, if we would identify architecture or even only postmodern architecture with that. It would lose all contact to the tradition of architecture as the art of spatial forms.

The examples that fit the conception of Jencks can be easily found in Venturi, Scott Brown, and Izenour's *Learning from Las Vegas*.[15] It is the *decorated shed* that meets the idea of Jencks. The term already suggests that there is a resignation on architectonic spatial design. Instead, there is a façade design that works with brand-symbols, characters, and illumination. Here architecture becomes commercial art.

Now in fact it is correct, that postmodern architecture implies the return of decoration, after its condemnation in modernism, from Loos to Bauhaus. At the same time one has to admit, that architecture also serves to stage – the staging of power, of religiosity, or of democracy and art. But at the same time this does not mean that it has to do without its own instruments for the benefit of symbols. And this also doesn't imply that decoration and instruments of staging have to be at any prize symbols or even metaphors. A flat roof doesn't "mean shelter and psychological protection," as Jencks suggests,[16] one can say at best that it suggests security, that is, it contributes to the creation of an atmosphere of security. Nor can we consider decoration as an architectonic metaphor. If for example the perception of a house suggests the impression of a face and this effect is amplified by decoration, then this decoration is not a metaphor for the face, but an articulation of the sight of a *face*. The discussion on metaphors appears to be the strongest, when it concerns the transfer of elements of style on modernist buildings. Because here, in the words of Jencks, it is the question of *transfer* of elements from one code to another – if one can consider architectonic styles as codes. An ionic column in a modern building is not a metaphor for something else – as the lion for the audacious, aggressive warrior – but a symbol, that creates along the conventions an atmosphere of festivity and solidity.[17] This distinction is important because in the first case the architect has to presuppose the readability of the metaphor by his public – that is in general education – while he can bear in the second case on the immediate – even though culturally encoded – impression.

Conclusion

The usage of metaphors in language has an appealing character. That is also the reason for the original use of metaphorical discourse in aesthetics. A metaphor is an uncommon use of words – it is a terminological innovation. It therefore normally endows a new, surprising view and thus implies a cognitive function. And it achieves an accentuation and in many cases an illustration of the discussed object that it puts in the context of another domain.

Metaphors are thus not only an ornament of speech, but, in fact, the dynamic of the development of a language is considerably influenced by the constitution of metaphors. That implies that metaphors are, strictly speaking, only metaphors in their first utilization. Through repetition their appeal quickly disappears and they are then only words like others. An example for such dead metaphors is in German *Blatt* – leaf, page. This word was originally coming from the realm of plants and blossoms, today it is used without any charm for almost anything that is flat: saw blade, rudder blade, a sheet of paper. The word atmosphere was in the eighteenth century a metaphor coming from meteorology, today it has basically two meanings: that is, *upper atmosphere* and *tempered space*, so there is a duty to explicate the relationship of the two.[18] Another destiny of metaphors is to become a way of speaking. One recognizes in it a certain strangeness, but without being able to understand it, still the way of saying lives on as an understandable concept. Examples are – I'm sorry but they work only in German – *Ich komme auf Schusters Rappen* or *den inneren Schweinehund überwinden.*[19]

These are all procedures inside language. If we are discussing architecture proper, talking about metaphors in architecture is indeed a metaphor. If metaphors concern expressions, which appear in the discourse about architecture, they are harmless and move inside the domain of the common. In that case they don't need any further justification, besides the warning, that metaphoric expression implicitly covers the incapacity, to speak directly about architecture wording its particularities. If one doesn't describe the elements of architecture only metaphorically, but instead calls these elements themselves metaphors, then the whole discourse is a metaphor. In fact, this discourse translates a word that actually corresponds to phenomena of language, on works of architecture. This is similar to Roland Barthes who has spoken of a language of fashion.[20] Thereby he has reconstructed the elements of industrial textile production as a system of signifiers that do not have a corresponding signified. A collar, a sleeve doesn't mean anything. But they could have, in the totality of a costume, a certain radiation. What Barthes has accomplished thereby is to reveal the modular structure of textile design. Something similar results from architectural semiotics. It is therefore not by chance, that it has been developed during the age of industrial and modular construction. But it is something different, when architecture becomes marketing, as Venturi and the others have represented it. In that case the elements of architecture have a meaning, they are signs. Considered in the overall this architectural thinking – and its corresponding way of designing – means decay. Architecture then only sets signs and renounces its original task that is to design and construct spaces.

Notes

1 Charles Jencks, *The Language of Post-Modern Architecture*, London, Academy Editions, 1977, p. 60.
2 See Chapter 13 in this book.
3 For this differentiation, see my articles "Leibliche Anwesenheit im Raum," in *Ästhetik und Kommunikation* 108, Mars 2000, pp. 67–76; and "Der Raum leiblicher Anwesenheit und der Raum als Medium von Darstellung," in Sybille Krämer (ed.), *Performativität und Medialität*, München, Wilhelm Fink, 2004, pp. 129–40.
4 Aristotle, *Poetics*, transl. Malcolm Heath, London, Penguin, 1996, p. 37.
5 According to W. Weinreich. See his article "Metapher," in *Historischen Wörterbuch der Philosophie*, Bd. 5, Stuttgart, Schabe, 1980, pp. 1179–86.
6 Aristotle, *Rhetoric*, III.4 1406b.
7 Homer, *Iliad*, XX, 158–77.
8 Precisely speaking Homer does not call Achilles a lion but says that he attacks Aeneas like a lion, λεων ως. A metaphor is also called an abbreviated comparison.
9 See George Lakoff and Mark Johnson, *Metaphors We Life by*. Chicago, Chicago University Press, 1980. For an overview of modern theories of metaphor, see Juliana Goschler in *Ihrem Buch Metaphern für das Gehirn. Eine kognitiv-linguistischen Studie*, Berlin, Frank & Timme, 2008.
10 See my article "Über die Physiognomie des Sokrates und Physiognomik überhaupt," in G. Böhme, *Der Typ Sokrates*, Frankfurt/M., Suhrkamp, stw, 3. Extended edn 2002, pp. 210–33.
11 Kant says: "It is one and the same spontaneity which in the one case, under the title of imagination, and in the other, under the title of understanding, brings combination into the manifold of intuition." Note to B 152. Immanuel Kant *Critique of Pure Reason*, trans. Norman Kemp Smith, New York, St. Martin's Press, 1965, p. 171f.
12 Jencks, *The Language of Post-Modern Architecture*, p. 40.
13 Ibid., p. 43.
14 For this concept, see my book *Architektur und Atmosphäre*, Wilhelm Fink Verlag, Munich, 2006.
15 Robert Venturi, Denise Scott-Brown, and Steven Izenour, *Learning from Las Vegas*, rev. edn, *The Forgotten Symbolism of Architectural Form*, Cambridge, MA, MIT Press, 1977.
16 Jencks, *The Language of Post-Modern Architecture*, p. 63.
17 Ibid., p. 55, has an example which through obviously being mere facade denies the emotional impression one may get.
18 See my article "Mir läuft ein Schauer übern ganzen Leib; das Wetter, die Witterungslehre und die Sprache der Gefühle," *in Goethe-Jahrbuch 124*, 2007, Göttingen, Wallstein, pp. 133–41.
19 The dictionary (Dr. Karl Wildhagen's *German–English Dictionary*, Wiesbaden, Brandstetter, 1972) says that *"auf Schusters Rappen kommen"* is equivalent to "on Shank's mare" and that *"den inneren Schweinehund überwinden"* is equivalent to "overcome one's baser feelings" or "the devil inside."
20 Roland Barthes, *Système de la mode*, Paris, Seuil, 1983.

Part IV
Light and sound

17 Acoustic atmospheres[1]

Introduction

Ever since aesthetic modernism, that is, since the days of Baudelaire, there has been an unremitting competition between the development of the arts and aesthetic theory. In the process, not only have developments in the visual arts, in their avant-garde movements, time and again transcended existing ideas of what art is, but developments in music have done so as well. *Musique concrète* and sound installations, in particular, forced a revision of music theory and, moreover, changes in the fundamental concepts of aesthetics in general. To outline the basic issues from the outset, these changes concern: the expansion of sound material, the notion of music as environment art, and the precedence of hearing.

Aesthetics was originally conceived, in the mid-eighteenth century, by Alexander Gottlieb Baumgarten, as a theory of sensory perception. However, it all too rapidly evolved into a theory of taste and limited itself, in its object of study, to artworks. Whereas, with Kant, aesthetics still seemed to be fundamentally an aesthetics of nature,[2] already, with Hegel, it had devolved into a mere preliminary to aesthetics proper, i.e. the theory of the work of art. Aesthetics, from then on, essentially served the development of aesthetic judgment and, by extension, art criticism and had completely abandoned the field of sensory experience and affective resonance.

The state of the aesthetics of atmospheres

The extraordinary advantage of an aesthetics of atmospheres is that it relies on a broad repository of everyday experiences. We speak of a "bright" valley, of an "oppressive atmosphere" before an impending thunderstorm, of the "tense atmosphere" of a meeting, and these are notions that are easily communicable. If atmospheres are moods pervading the air, the phenomenon thus described is familiar to all of us. What is more, there are almost endless ways and means of speaking about atmospheres, of characterizing atmospheres. We speak of a "serious atmosphere," an "ominous atmosphere," but we also speak of an "atmosphere of violence," or a "holy atmosphere," and we even speak of "boudoir atmosphere," of a "petit-bourgeois atmosphere," of the "atmosphere of the 1920s."

Based on these everyday experiences and ways of speaking, the term atmosphere has meanwhile been developed into a scientific concept.[3] What is special about this concept and what, at the same time, makes it theoretically problematic is that it denotes an intermediate phenomenon. Atmospheres are something in between subject and object: they can be defined as more or less objective sentiments indeterminately projected into the environment. In the same way, however, they have to be defined as subjective insofar as they are predicated on a subject experiencing them. Yet it is precisely in this intermediacy that its great value lies, for through it is united what traditionally has been separate: as an aesthetics of production on the one hand and an aesthetics of reception on the other. To be sure, atmospheres can be created and there are elaborate arts that deal specifically with the creation of atmospheres. They revolve around the deployment of eminently material, technical devices, however, not as causal agents, but rather as generators of atmospheres. The art of set design is paradigmatic for this approach to atmospheres. On the other hand, atmospheres are experienced in a state of affective resonance and we can only tell what their nature is by exposing ourselves to them by being there physically, in order to perceive them in our particular frame of mind. This is the classical aspect of the aesthetics of reception.

For the study of atmospheres, both the contrastive and the ingressive forms of experience have been proposed: the specificity of atmospheres is most readily experienced when their respective nature stands out, i.e. not when they have already descended into unobtrusiveness as something evenly surrounding us. Consequently, they are experienced through contrast, that is, when finding ourselves in atmospheres that clash with our own emotional state, or when entering into them by moving from one atmosphere to another. In this case, atmospheres are experienced as *anmutungen*,[4] impressions that tend to affect a particular mood.

On the side of the aesthetics of production, atmospheres are examined, as mentioned earlier, with regard to what creates them: objects, their features, arrangements, light, sound, etc. What is crucial, however, particularly for the ontology of the object, is that it is not the properties a thing possesses – that define it and distinguish it from others – that matter here, but rather the properties through which it radiates into space. The point is, more precisely, to read properties as "ecstasies,"[5] i.e. as ways in which an object comes out of its shell, as it were, and transforms the sphere of its presence. The study of ecstasies is of particular importance for the arts of design and set design, since the focus here is less on the objective properties and functions of things than on their dramatic value.

The aesthetics of atmospheres which, hence, has its roots in ecological aesthetics, is in the meantime evolving into a rehabilitation of the original approach of Alexander Gottlieb Baumgarten, i.e. of aesthetics as *aisthetik*, as a general theory of perception. It has by now demonstrated its revelatory power in a series of case studies, for example, on the atmosphere of a city, on light as atmosphere, on the atmospherics of twilight, on the atmosphere of church interiors, on music as atmosphere,[6] and, finally, in the study of the atmospheres of human interaction.[7]

The aesthetic conquest of acoustic space

Since the ancient Greeks, music as an artistic doctrine has been understood as the knowledge of tones, with the nature of tones being determined by harmonious distances to a fundamental, the intervals. To us today, this concept of music seems limited beyond comprehension. In comparison, the twentieth century has brought with it an immense, multifariously unfolding expansion of the material of music. One can quite aptly refer to it as a conquest of acoustic space. From tonality through chromaticism, things have proceeded to a gradual expansion of the acoustic material allowed in music, eventually including pure sound and noise. It started out with intermediate tones in the intervals of chromaticism and with how tones as such are created and made up – the sounding, pulling, striking – and with growing interest in the individual character of instruments, in their particular voices, sound gained in importance as well. Then, with the perversion of instruments by striking and scratching their resonators, as well as through ever newer percussion instruments, a wealth of sounds beyond mere tones were admitted in music. Finally, through tape recordings, everyday sounds, street scenes, nature sounds, and the acoustic world of the factory were introduced into music. Today, the technique of sampling makes any kind of acoustic material available for musical compositions.

In addition to this expansion of the material of music, a fundamental change in – or again, more precisely, an expansion of – the nature of music can be observed as well. Well into the twentieth century, the dogma prevailed that music is a temporal art, finding its true essence in its temporal form, in the moment-transcending unity of the musical action. From the basic demand for cadence and a return to the tonic through melody and theme, the structure of movements to the unity of the symphony: the essence of music lay in the sequence bound into unity. Even in the 12-tone music of, say, Schoenberg, musicality as such was embedded in the unity of the sequence. This understanding of music did not become obsolete, but it was relativized insofar as music was newly discovered as an environment art and, in New Music, was more or less explicitly developed as such. That music fills spaces and that space, through resonance and reverberation, constitutes an integral part of its effect was something that had been known all along. What was newly realized was that the individual tone, the ensemble of tones, as well as the sequence of tones – or, more precisely, the sequence of sounds – have spatial shapes, forming figures and ensembles in space. This was previously not a matter that music reflected on. Most likely, it was only the modern electronic techniques of reproducing and producing music that made this field manageable and, consequently, also drew attention to it. To have a sound buzz like an insect through space or, perhaps, have it rise above a muffled sound mass and explode like fireworks – these are possibilities that only technology has provided, thereby drawing attention to something that, in certain respects, has always been a part of music. Even the Greek words for high and low, *oxýs* and *barýs*, meaning "sharp," "acid," and "heavy," "broad," point to this fact. New Music, however, started to make very deliberate use of the spatial shape of music, partly by returning to the classical instruments, partly through electronic installations, thus creating awareness of space as an essential dimension of musical composition in the first place. In some cases, this dimension can turn into the actual dimension of a musical artwork which, consequently, can no longer be expected to have something like a beginning or an end or, for that matter, a principle of time-transcending form.

Almost all the characteristics mentioned of music expanded in terms of its sound material and of music as environment art can be found in the work of Hans Peter Kuhn. His installations are, moreover, intermedially expanded in the sense that they integrate sound and light, creating environments of unsettling, yet also liberating, artificiality. On the part of the sound, this artificiality results from the fact that Kuhn not only samples the original sounds which come from instrumental music, nature, and the worlds of manufacturing and traffic, but as such also electronically processes them. On the part of the light, it is the result of Kuhn not hiding, but, on the contrary, practically parading the fact that the play of light in his works is created by artificial light. Therefore, one could say that Kuhn almost inverts the dictum of Kantian aesthetics according to which art should imitate nature, in order to be beautiful: all the things of nature used, from the materials and the sounds to the light, have to appear artificial, in order to be … what? Beautiful? Yes, that too, but that's not the point: the point is to be *aesthetic*. And being aesthetic means, according to the new aesthetics, that the work manifests itself to the individual who experiences it in such a way that its manifestation as such becomes evident as well. Art, in this case, becomes a demonstration of the possibility of creating atmospheres.

It is precisely its tendency toward environment art that has moved music into the realm of an aesthetics of atmospheres. The environments or spaces at issue here cannot, in fact, simply be identified with geometric space – at best, if at all, they can be identified with topological space. To be sure, there are, in musical space, directions, there is something resembling form and a kind of beside each other as well, but all this not in the strict sense of a separation, but rather as changing, merging, emerging, and disappearing shapes. Moreover, this space is experienced affectively. That which is expansive is experienced as heavy and oppressive, that which is rising as relieving and joyous, that which is shattering as jolly, etc. Taken together, one realizes that musical space is, in fact, the expanded space of the body, i.e. the sensing out into space, which is shaped and articulated by music.

This realization that music is the fundamental atmospheric art form has solved an old, always bothersome and yet unavoidable problem for music theory: the question what the so-called emotional effect of music actually consists in. As opposed to the inadequate theories of association or the theories calling upon imagination as an intermediate link, the aesthetics of atmospheres can provide the simple answer that music as such is the transformation of physically sensed space. Music shapes the way the listener finds him- or herself in space, it intervenes directly in his or her physical economy. Practitioners have made use of this long before any theoretical realization: already with silent movies, music served to lend the image three-dimensionality as well as emotional depth. Film music later followed this practice. With regard to radio plays or radio features people actually refer to an "atmos" serving as background to the narrative: music or, more generally, acoustic action to provide the spoken words with atmosphere. Similarly, in bars, an atmosphere is created through a specific sound, and muzak is used to make it pleasant to be at airports, in subway stations, or at the dentist's, or to brighten and enliven one's presence in department stores or hotel lobbies.

If twentieth-century music has expanded acoustic space by introducing new sound materials including technical sounds and everyday samples and even noise, and if music finally has been transformed from a temporal art form into an art form that deliberately shapes affective environments, it has benefited in this conquest of space from an entirely different development. I am referring to the World Soundscape Project founded in the 1970s by Murray Schafer.[8] Here, the world of natural sounds, the acoustic life of a city, the acoustic nature of technology and work were researched, documented and, eventually, the newly mined material was used for compositions. Acoustics and sound engineers collaborated with musicians or created works themselves as composers. Something that, from the point of view of music, was a development toward the expansion of musical materials presented itself in this respect as a discovery of the musicality of the world itself. Certainly, it had always been realized that birds or, for that matter, whales have their own music. But this was about more: about the discovery of acoustic characters, or rather, of the characteristic shape of habitats, be they natural ones, such as the sea, the forest, and other landscapes, or the habitats of cities and towns. It turned out that documenting these acoustic worlds alone required a process of distillation and composition, in order to make them accessible to those not at home in these parts. What could be more natural than turning this process of distillation and composition into an explicit process of creation, thereby partly meeting music, partly merging with it? John Cage's work, *Roaratorio*, can be considered exemplary of the latter.

In the meantime, it has been discovered that a sense of home is essentially conveyed by the sound of a region and that the specific feeling of a lifestyle, of an urban or rural atmosphere, is quite fundamentally determined by the particular acoustic environment. This means that, today, what constitutes a landscape can no longer be reduced to things visible and that urban planning, for instance, can no longer merely devote itself to reducing or protecting against noise, but rather has to attend to the particular character of the acoustic atmospheres of squares, pedestrian zones, and entire cities.

These two developments, i.e. of twentieth-century music and of the Soundscape Project, should not be considered independently of the development of technology. Just as the development of music as environment art is hardly conceivable without electronic reproduction and production technology, the examination of acoustic landscapes is inconceivable without electronic recording and reproduction technology. The development of acoustic engineering in the twentieth century, however, has had yet another effect entirely independent of the two mentioned: the omnipresence of music. Music which, in the past millennia of the European tradition, was something related to celebrations and special occasions, has become a cheap common consumer commodity. Through radio broadcasting and television, music is available any time, and through the acoustic furnishing of public spaces our acoustic environments tend to be already occupied by, or at least interspersed with, music.

What are the consequences of this development? Based on the latter observation, one could speak of an acoustic pollution of our environment.[9] On the other hand, however, we have to acknowledge that the acoustic awareness of the average person has evolved significantly, not only in the sense that musical needs and, as part of these, acoustic demands have increased considerably, but also in the sense that, for the general public, hearing as such has become a dimension of life and an area of gratification. Of course, it cannot be denied that the noise of the modern world and the invasion of public space by music have, at the same time, resulted in a common practice of tuning out. Nevertheless, hearing has indeed developed from an instrumental sense – hearing something – into a way of taking part in the life of the world. As far as music is concerned, the developments mentioned have obviously resulted in blurring its boundaries. Whereas, at the outset of the history of Western music, it had actually defined itself by establishing such boundaries, the continuous expansion of its field has tended to render any boundaries indefinite. When Thierry de Duve[10] asserted with regard to the visual arts that, after Duchamp, the fundamental question of aesthetics has shifted from "What is beautiful?" to "What is art?," the same holds true for music.

Acoustic atmospheres

Perhaps, by pointing to acoustic atmospheres this question has received a provisional answer, "provisional" meaning: an answer defining what, for our times, is characteristic of musical experience. It is to be expected that in different times after us, perhaps soon after us, other answers will have to be provided. What is certain, however, is that the great age of Platonism in music has come to an end. Plato criticized those who used their ears to try to find out what harmonic intervals are.[11] And even Adorno could still claim that the proper way to listen to a symphony is to read the score. How infinitely far away are we from that today! With modern music, it has become doubtful to us whether it can still be adequately written down. It appears that sensuality in music has undergone rehabilitation and that we have to insist, against the entire Platonic age, that music is something that can be experienced only through hearing. We may even have to argue that the real subject of music is hearing itself. Modern art in general has been said to be self-reflective, to turn to art itself, to its social status, its anthropological meaning, and its sheer appearance for its subject matter. In the visual arts, this turn toward self-reflection had a clear and identifiable purpose. In many instances, works created in the visual arts were no longer about representing something, but rather about the experience of seeing itself. Starting, perhaps, as early as Turner or the Impressionists, this became particularly obvious with artists such as Joseph Albers, Barnett Newman, and Mark Rothko. In music, this development may not have become as evident, being, in certain respects, much more in the nature of things. For it has always been clear that music, as opposed to images, is non-representational, does not represent anything. To be sure, there has been painterly music, there has been program music. In reality, however, these were undeniably deviations, with music tending to serve a different cause. Kant already suggested that music is the language of emotion, a statement that, following the common semiotic view of language, could readily be understood to mean that music denotes emotions, i.e. is representational. But that is not what Kant meant, since he actually distinguished the tone in which something is said from the meaning communicated through signs, the tone allowing us to share directly in the emotion of the speaker.[12] Music, to him, was the independent form this manner of communicating emotions took. Today, we have reason to generalize this idea: accordingly, what distinguishes music is that it is about acoustic atmospheres themselves. Music's boundaries would thus be marked out very differently from the way they presented themselves to us in the Platonic tradition. Music, in this tradition, was essentially defined by a limitation of sound materials, or more precisely, by a restriction of the acoustic space through which musical tones were defined. Today, we can argue that we are dealing with music whenever an acoustic event is about the acoustic atmosphere as such, i.e. is about hearing as such, rather than about hearing something. This requires further clarification, but one thing we can already say is that, as a result, music most certainly does not have to be man-made anymore.

What does it mean, then, that it is about hearing as such, rather than about hearing something? What we first realize when asking this question is the great extent to which hearing is usually object-related. "I hear a car passing by," "I hear the clock strike 12," "I hear someone talking," "I hear a mosquito," "I hear the horn of a ship." This way of hearing is useful and plausible; it serves us to identify objects and their position in space. But, in certain respects, in this mode of hearing, hearing as such goes unnoticed. Of course, instead of saying: "I hear a dog barking," one could say: "I hear the barking of a dog." But in this case we would, in fact, be dealing with a different kind of hearing. The barking, to be sure, belongs to the dog. It is "one" mode of its being in space. However, what distinguishes voices, tones, and sounds is that they can be separated from their sources or can separate themselves from them, filling the space and traveling across it almost like objects. Perceiving acoustic phenomena in this way – i.e., perceiving them as such and not as ways in which things express themselves – requires a change of focus. As children of our times we have actually oftentimes practiced this through the use of acoustic equipment, in particular, by listening through headphones. To many of us it has been humiliating that this is what it took to finally realize that acoustic environments are something in themselves, independent of the world of objects and not identical with actual space. But, of course, acoustic space is experienced in real space, as well. It is, however, the space of the body, the space of my own presence, which is expanded through the reach of physical perception. In hearing that tone, voice, and sound do not transfer to the objects from which they may have originated, the listener perceives voice, tone, and sound as a transformation of the space of his own presence. Whoever listens in this manner is dangerously exposed, letting himself out into wide open space, and therefore susceptible to being struck by acoustic events. Sweet melodies may carry him away, claps of thunder may strike him down, buzzing sounds may be threatening to him, a biting sound may injure him. Hearing is a form of being beside oneself and, for this very reason, it can provide a gratifying awareness of being in the world.

These are things one has to experience; they cannot be conveyed verbally. There is, however, a suitable analogy that can clarify what is meant. A philosopher essentially thinking in mechanistic terms, Descartes, nevertheless, already inquired about the place where someone poking at a stone with a stick actually perceives this stone. His answer was, as it would be later, in twentieth-century gestalt psychology that one perceives the stone in the place where it is. This has been described as *einleibung*, as physically assimilating, the stick, which, indeed, is not entirely wrong. Strictly speaking, however, we are dealing with an expansion of the space of the body. How much more pertinent even than to poking with a stick is it to argue with regard to hearing that, in hearing, we are "outside"? And this being outside does not encounter voices, tones, and sounds out there, but, rather, is itself shaped, moved, modeled, carved, cut, raised, pressed down, expanded and restricted by voices, sounds, noises.

The best model of hearing that has been around until now asserts that what is heard is simultaneously reconstituted internally. This was the resonance model of hearing and it had some plausibility in the familiar experience that we tend to sing along in our heads with a tune we hear. This model, however, suffers from the untenable topology of an inside and outside and quickly finds its limits, particularly, vis-à-vis the complexity and unfamiliarity of things heard. After all, we would hardly be able to sing along in our heads with the sounds of an engine room with all its whirring, shrilling, whistling, and buzzing. We actually do not hear these sounds in our heads at all, but rather outside. What is resonating here, and the space in which these voices, tones, and sounds occur, is the space of the body itself. To be sure, this experience probably occurs only rarely, or more precisely, only rarely in a pure form, since, in a way, it underlies any hearing experience. It is just that the self does not usually abandon itself to hearing and preserves itself by relegating the voices, tones, and sounds to their sources, thus skipping the experience of the in-between.

Conclusion

In conclusion, we should return once more to the beginning which, as we remember, consisted of an ecological question. The development of music in the twentieth century has resulted in music itself becoming an element of the environment. Its functionalization as a factor of interior design – it is common to speak of acoustic design – has, to a certain extent, reduced it to the atmospheric dimension. At the same time, however, avant-garde music on the one hand and the World Soundscape Project on the other have virtually elevated the acoustic atmosphere to music's essence. In the process, the voices of things and the concert of the world have received growing attention and hearing has gained in importance to life. When we add this up, ecological aesthetics in the realm of the acoustic is not merely a complement to scientific ecology, but rather has its own purpose: the awareness, the preservation, and the shaping of acoustic space. Accordingly, the question as to what constitutes a humane environment poses itself as an inquiry into the particular characters of acoustic atmospheres. Here, too, it is essential to transcend the merely scientific approach which can, at best, register noise in terms of decibels, and to ask what acoustic characters the spaces we live in should have.

Notes

1 Translated by Norbert Ruebsaat.
2 This impression dissolves on closer scrutiny. As it turns out, it focuses precisely on examples from the realm of design. Gernot Böhme, "Index over de aestetiske exempler in Kants 'Kritik der Urteilskraft,'" *Kritik*, 105, 1993, 79–80 and *Kants Kritik der Urteilskraft in neuer Sicht*. Frankfurt/M., Suhrkamp, 1999.
3 Gernot Böhme, *Atmosphäre. Essays zur neuen Ästhetik*, 1995, Frankfurt/M., Suhrkamp, 2nd edn, 1997; Michael Hauskeller, *Atmosphären erleben. Philosophische Untersuchungen zur Sinneswahrnehmung*, Berlin, Akademie Verlag, 1995.
4 Gernot Böhme, *Anmutungen. Über das Atmosphärische*, Ostfildern, Tertium, 1998.
5 On this concept, see Böhme, *Atmosphäre*, Part III.
6 These studies can be found in Böhme, *Anmutungen*.
7 G. Böhme, "Kommunikative Atmosphären," in T. Arncken, D. Rapp, and H.-C. Zehnter (eds.), *Eine Rose für Jochen Bockemühl. Special issue of Elemente der Naturwissenschaft*, Dürnau, Kooperative, 1999. For the other studies, see Böhme, *Anmutungen*.
8 A more recent example of R. Murray Schafer's vast body of work is *Voices of Tyranny – Temples of Silence*, Indian River, Ont., Arcana Editions, 1993.
9 For a critical look at this, see Hildegard Westerkamp, "Listening and Soundmaking: A Study of Music-as-Environment," in Dan Lander and Micah Lescier (eds.), *Sound by Artists*, Toronto, Art Metropole, 1990, pp. 227–34.
10 Thierry de Duve, *Kant after Duchamp*, Cambridge, MA, MIT Press, 1996.
11 Platon, *Politeia*, VII, 53 1a.
12 Immanuel Kant, *Kritik der Urteilskraft*, 1799, p. 219.

18 Music and architecture[1]

The trivial connections

There are trivial connections between architecture and music that have been much abused and that can still be evoked again and again. More precisely, they take two forms. First, there is the metaphorical connection between architecture and music. Thus one speaks of the architecture of a Bach fugue, and conversely one can say that the architecture of the Alhambra in Granada is like that of a Bach fugue. Such metaphors can become embarrassing, namely when one speaks about one field with terms borrowed from the other. Sometimes, however, there is indeed a deeper truth to it. In the example just given, for example, there is a structural relationship: the modulation, permutation, and mirroring of a basic pattern.

The other trivial relationship between music and architecture is that music, as a rule, is performed in built spaces, and then of course the acoustic qualities of the spaces – what is musically possible within them – are highly significant Acoustics as a specialized discipline within physics is thus an important science as an aid to architecture to the extent the latter is concerned with buildings for acoustic performance, that is, for excellent tonal qualities, for intelligibility, and so on. The connection here is generally one-sided, that is, architects are asked to create spaces in which it is possible to present music of good quality. It is therefore a very welcome development that musicians – or better, composers – are now also interpreting the relationship from the other direction, namely by creating works for existing spaces that suit the acoustics of these rooms in particular ways and thus, as Gerhard Müller-Hornbach says, conceive the architectural space as a kind of extended instrument for which they can compose. Nevertheless, I would say that this too is just another trivial connection between architecture and music, in that this approach understands space as a given of physics, with particular attention paid to its acoustic qualities. It does not yet claim that music creates spaces just as architecture does and that therefore they could meet in the middle, namely around the theme of space.

The center: the space

The true, non-trivial relationship between architecture and music becomes a theme when music is conceived as an art of space, just as architecture is an art of space. Astonishingly, this conception continues to be non-trivial. Many consider it a challenging thesis that not only allows existing things, in music as well as architecture, to be read in new ways but also contains great innovative potential for the future.

First, the thesis: music is an art of space. This thesis stands in contrast to one traditional conception of music, which states that music is essentially an art of time. The renowned music theorist Carl Dahlhaus summed up this view when he said that music is a "shaping of time" or "shaped time."[2] This view is not, of course, wrong; on the contrary, it must be said that it addresses a very important dimension of music. It becomes wrong, however, at least from a historical perspective, if it is understood as a definition of the essence of music in general. That the essential aspect of music consists in bringing one after another of a sequence of notes into a unity – that is, to give it a form – is a characterization that certainly applies to large periods within the history of European music, but it no longer applies to the evolution of New Music in particular; nor does it apply to other types of music that have since become part of the international music scene, especially those deriving from Asia. What these other directions have abandoned is the requirement that a piece of music have a beginning and an end determined by its inner structure – traditionally, this was achieved by the harmonic arrangement, by tonic and cadence – and in general that there be a defined order to the sequence that is necessary to understand the music. In the meanwhile, there are compositions in which the musical event consists in the manifestation of individual notes, noises, or configurations against an atmospheric background or silence; pieces that fill the space in such a way that the listener can experience different sequences by moving through the space (Gerhard Müller-Hornbach); and finally pieces whose open-endedness is essential (Claus-Steffen Mahnkopf, *Hommage à Thomas Pynchon*). This is just to name some examples in which composers have abandoned the dispositive aspect of shaping time. Conversely, developments in New Music, in particular the introduction of electronic means of production and reproduction, have also made space itself a theme for musical composition. As soon as one has said that, and has thus focused on what is innovative about this perspective, there is a risk of seeing what is new in the thesis "music is an art of space" as something old and familiar. Performing musicians have, of course, always related to spaces, and have had to take into account the reverberation, echo, and resonance of various spaces. There are even early examples of the structure of the space already playing a role at the composition stage, namely baroque music composed for choirs distributed within the church. These always counted on existing spaces, however, and did not yet have the awareness that music creates spaces or shapes spaces. This innovative idea seems to have been first conceived by Marcel Duchamp, who wrote on a slip of paper in 1913 that it was necessary to create acoustic sculpture. This composition by Marcel Duchamp was performed by the Ensemble Recherche at a Duchamp Symposium in Darmstadt in 2001. It would perhaps be more accurate to say they executed the piece, and in a form that was still inadequate, because in essence the ensemble attempted to realize Duchamp's idea by distributing the musicians in the room. That is certainly not wrong, but it raises the question – one that can only be answered empirically – whether the result was anything like an acoustic sculpture. What is or would be an acoustic sculpture? To answer it, it is necessary to refer to certain well-known but usually subliminal experiences of listening to music or to traditional figures of speech that have hardly been taken seriously in their literal sense. Consider, for example, the fact that notes are described as high and low or, in ancient Greek, barys and oxys. High and low are spatial references, but they have no justification in the spatial arrangement of notes or finger-rings on instruments. Rather, these expressions clearly refer to the way the notes are truly heard. This is even clearer in the Greek expressions barys and oxys. Barys means "heavy," "flat"; oxys, "pointed," "sharp." These expressions are clearly expressions of a spatial experience of sound, or at least a synesthetic one. Following these linguistic clues, we discover that we certainly do hear notes, musical configurations, and music in general spatially, that is, neither radiating from spatially distributed sources nor simply filling the space but rather as a presence with a specific, changing spatial shape. Notes and musical configurations can rise and fall; they can blaze; they can be sharp or flat; they can be torn; can move around in space; they can be cloudy or pointillist; and in general they can present themselves in many spatial forms. That spatial form has become a theme both in the production and in the reception of music is probably due to the development of technological means for the production and reception of music. On the one hand, listening with headphones has made something like presence and perception in a purely acoustic space possible – that is, in a space without objects and objective structures; on the other hand, with the advent of sampling and the coordination of loudspeakers distributed in the room using computer software, it has become possible to control the movement of sound and the forming of acoustic shapes in space.

Astonishingly, the thesis that architecture is a spatial art often has a non-trivial sense and contains a potential for innovation. That architecture has to do with space and that the shaping of space is essential to it can surely meet with unanimous agreement. The problem is simply that the traditional understanding of architecture as spatial art has been dominated by the traditional, and still dominant, European ontology that goes back to Aristotle. This ontology conceives all that exists within the scheme of form and material. In this view, architecture is essentially the shaping of masses – buildings are thus large, walk-in sculptures. By contrast, during the twentieth century, slowly but surely, an understanding of architecture evolved in which architecture is essentially the shaping of spaces, if not the creation of spaces. Characteristic of this view is Peter Zumthor's formulation that architecture has two basic kinds of spatial design: on the one hand, the delimitation of a space within the architectonic body and, on the other, the enclosing of an area associated with infinite space by means of an open body.[3] He is clearly thinking of examples like a hall, on the one hand, and a loggia or square, on the other. That is, of course, only the beginning, and Zumthor's alternatives should not be viewed as a complete disjunction. Indeed, even the second possibility he mentions could be read the other way around as well. An open enclosure – say, Saint Peter's Square in Rome or the semicircular garden of Schwetzingen Castle – does not just create a space, as Peter Zumthor himself indicates, it also points to the openness, that is, it opens up the space for us to experience precisely by means of its broadness. If we follow this thought, or more precisely, the corresponding experiences, it opens up a whole spectrum of other architectonic possibilities for shaping space. What does a medieval fortress at the peak of a mountain do with the space? What happens to space when a dog barks in the distance, or a plane is seen in the sky? What spatial experience did Jonathan Borofsky's *Man Walking to the Sky* evoke in front of the Fridericianum in Kassel?[4] Not all of these are architectonic examples, but what happens in them can certainly be achieved by architectonic means. A fortress on a mountain peak or hilltop concentrates the space. When we are in the corresponding landscape, we certainly sense the centering and condensation of space in the vicinity of the fortress. A dog barking in the distance or an airplane in the sky articulates the distance. The experience of the broadness of the space is conveyed precisely by the dot-like thing within that expanse. Jonathan Borofsky's sculpture at *documenta 9* conveyed a visible suggestion of movement. This suggestion of movement, of walking to the sky, would also have been conveyed by the diagonally placed beams alone, the artist's achievement merely consisted in drawing attention to this suggestion of movement. The communication of suggestions of movement, the concentration of space, the opening of an expanse of space by means of articulation – these are all methods of treating space that have in principle always been part of architecture. At most, one can say that the architecture of the second half of the twentieth century began to make this treatment of space its true theme. Naturally, on the one hand, new building materials, from steel construction by way of reinforced concrete and Plexiglas to plastic, contributed a great deal to this; on the other hand, however, so did the flourishing development of capitalism, that is, its transformation into the aesthetic economy.[5] In summary, we can say that space is no longer a given for architecture, that rather the task of architecture is essentially the creation and shaping of spaces and experiences of space. But what sort of spaces are these and how do we understand the term space when we use it in this way?

Atmospheres

European thinking about space is familiar with, in essence, two traditions. One, which goes back to Aristotle, thinks of space as topos, as place. In mathematics, topology is the study of relationships of position, interrelations, and surroundings. There is no measuring in topology. The other spatial concept is space as spatium, that is, as distance and intervals. It goes back to René Descartes. Seen mathematically, such a space is essentially determined metrically, that is, by the fact that the spatial relations can be grasped quantitatively. Both spatial concepts have in common that they deal with spaces in which there are bodies and that they are considered in relation to bodies. In distinction to these concepts of space, there is the space of bodily presence.[6] This is the space that we experience through our spatial presence, that is, the space that we feel bodily or with our own body. This space is essentially constituted by narrowness and broadness. It is, in contrast to the two mathematical types of space mentioned above, anisotropic, that is, it is centered, determined by an absolute here at which I am located. The possibilities for the musical and the architectonic designing of space – ideas that I mentioned in the second section of this chapter – essentially refer to the space of bodily presence. The spatial structures under discussion are essentially spatial structures for people who are experiencing things: for someone listening to music or someone bodily present in a building. Thus it is clear that the developments in both music and architecture – the developments that make it possible to characterize both as arts of space – are of the sort that essentially relate to people in terms of their bodily presence, that is, that take into account the possibilities of the experience of space as an essential aspect of composition or construction.

Bodily space is neither the place a person's body takes up nor the volume that it constitutes. A person's bodily space is the sphere of his or her material presence. The latter continually transcends the limits of the body. This is best made clear by an example that dates back to Descartes: a blind man who is feeling the ground with his cane and notices a pebble does not sense the pebble on the surface of his body – on the skin of the hand holding the cane – but rather where it in fact lies: on the ground, at the end of his cane. Analogously, in architecture the sense of narrowness and broadness in a given space is not a sense of the body's narrowness or broadness but rather how much a sense of the feeling our way out into the space is constricted or expanded. And in music?

Listening to music through headphones clearly demonstrates that we are within the thing that we perceive acoustically. There are people who in a physical – more precisely, neurophysiological – way conceive of listening through headphones as music in the head, but that is only because, looking from outside, so to speak, at the person listening as an experimental subject, they have to locate the sounds that they hear somewhere in physical space. The person listening, however, hears the music outside and perceives it as being in a room filled with music. It is very important that this form of listening does not differ from listening without headphones. This experiment demonstrates that the space of listening is a space of bodily presence that is independent of the existence of concrete things. When listening, listeners are in some sense outside. They sense the broadness, and this broadness is articulated and shaped in a certain way by the music that they perceive. Bodily space – that is, the space we experience only through our bodily presence – obtains its character not only through constriction and expansion and not only through direction, centering, concentration, and articulation. Rather, it always has an emotional character as well. Narrowness and broadness when experienced bodily are not, of course, emotionally neutral but have an effect on our frame of mind as well. In general, we might say that through our feeling (*Befindlichkeit*) we feel the character of the space in which we find ourselves (*sich finden*). A space, whether it is one shaped architectonically or musically, strikes us in a particular way. We say that the space has an atmosphere. This keyword opens up far-reaching perspectives for architecture and music, not only for the way they are understood but also for how they are practiced.[7] There is a positive and a critical side. The positive one is that one learns to understand better the emotional effect of both music and architecture. The attempt to understand music's effect on the emotions has hitherto been based, more or less helplessly, on associative theories or by reference to the painterly character of music. If, however, we assume that music modifies the space of bodily presence, then it is also immediately obvious that it modifies our feeling in space as well; and the same applies to architecture. As a rule, architecture is conceived of as a visual art. If architecture is rather understood as an art of space, its true experiential means is bodily feeling, by means of which the architectonic shaping of spaces has an immediate effect on our feeling. The same theory can, however, also be used as a critical instrument, in particular, to criticize the emotional manipulation of people that occurs when music is used as acoustic furniture – in department stores, subway stations, at the dentist – or the architecture of authoritarian buildings, such as churches or court buildings, that put visitors into a particular frame of mind and thus establish a certain disposition of activity and passivity. For music and architecture are not simply means to enhance and fulfill our lives, but are also instruments of power.

Notes

1 Translated by Steven Lindberg.
2 Carl Dahlhaus, *Esthetics of Music*, trans. by William W. Austin, Cambridge, UK, Cambridge University Press, 1982, p. 75 (German original, *Musikästhetik*, Cologne, 1967, p. 113).
3 Peter Zumthor, *Thinking Architecture*, Basle, Birkhäuser, 1999.
4 Its present location, an unfortunate choice, is in front of Kassel's old train station. See Figure 13.1, in this book, p. 139.
5 On this, see my essay "Contribution to the Critique of the Aesthetic Economy," *Thesis Eleven*, 73, 2003, 71–82 (German original, "Zur Kritik des ästhetischen Ökonomie," *Zeitschrift für kritische Theorie*, 12, 2001, 69–82).
6 On this, see Hermann Schmitz, "Der leibliche Raum," vol. 3, pt. 1 of *System der Philosophie*, Bonn, Bouvier, 1967. See my essay "The Space of Bodily Presence and Space as a Medium of Representation," www.ifs.tu-darmstadt.de/fileadmin/gradkoll/Publikationen/space-folder/pdf/Boehme.pdf; last accessed May 20, 2016.
7 See Gernot Böhme, *Anmutungen: Über das Atmosphärische*, Ostfildern, Edition Tertium, 1998 and Chapters 13 and 17 in this book (German original, "Atmosphären als Gegenstand der Architektur," in Philip Ursprung (ed.), *Herzog & de Meuron: Naturgeschichte*, Baden, Lars Müller, 2002, pp. 410–17).

19 The great concert of the world

Introduction

Residing in the archives of our cultural history is a theory of music which merits re-examining in view of developments in modern music since Schoenberg: the theory of the philosopher and mystic Jakob Böhme (1575–1624). It is to be found in his treatise *De signatura rerum*.[1] In this text Jakob Böhme conceives of things – more precisely, of everything that is – on the model of the musical instrument. The body is understood as a resonance body, the shape and materiality of which are responsible – as "tuning," which Böhme calls *signatura* – for the characteristic way in which a thing can express itself. At rest within the thing is its essence, *essentia*, which needs to be excited in order to manifest itself. On a large scale, Böhme makes God responsible for this, but in particular cases it may be another thing or a human being which, by blowing on it, causes a thing to sound.

What is crucial here is that Böhme has a theory of understanding according to which we understand an utterance if it causes an "inner bell" in us to vibrate. That is to say, understanding is resonance. In this way, what we call interaction becomes, for Böhme, a phenomenon of resonance. Things do not act on one another – as Descartes later conceived – by pressure and thrust, but by communication. The interconnectedness of the world therefore presents itself to him as a great concert. Might it be the case that what we call music is a part of this great concert, or our way of participating in that concert?

The art of modernism and the aesthetics of atmospheres

Since the inauguration of aesthetic modernism, that is, roughly since the time of Baudelaire, there has been a continuous race between developments in art and in aesthetic theory. The developments in art not only concern the visual arts, which in their avant-garde form have constantly overstepped notions of what art actually is, but also the field of music. Most especially, *musique concrète* and sound installations have enforced a revision of music theory and, more generally, have led to changes in the basic concepts of aesthetics. These changes, to sum it up straight away, concern the expansion of the sound material, the concept of music as a spatial art, the primacy of hearing and the return of voice.

Aesthetics was originally conceived, in the mid-eighteenth century by Alexander Gottlieb Baumgarten, as a theory of sensory cognition. All too quickly, however, it developed into a theory of taste and restricted its subject matter to works of art. Whereas in Kant aesthetics still seemed to be fundamentally an aesthetics of nature,[2] in Hegel nature was merely the antechamber of aesthetics proper, of the theory of the work of art. From that time on aesthetics primarily served aesthetic judgment and therefore art criticism, and had entirely vacated the field of sensuous experience and affectivity. For this reason, it has proved entirely incapable of comprehending the developments of modern art after Schoenberg and Duchamp. This becomes clear in the aesthetics of Adorno, who stands, as it were, on the threshold: he was unable to recognize or acknowledge the art character of jazz.

Since then a new aesthetics has developed, at the center of which is the concept of atmosphere. The extraordinary advantage of this aesthetics of atmospheres is that it is able to connect to a large reservoir of everyday experiences. One speaks of a "blissful landscape," of an "oppressive thundery mood" or the "tense atmosphere of a meeting," and one's meaning is readily understood. If atmospheres are understood as moods present in the air, a phenomenon has been registered which is familiar to everyone. What is more, there is a practically inexhaustible fund of expressions with which we talk about and characterize atmospheres. One speaks of a "serious atmosphere," a "threatening atmosphere," or a "sublime atmosphere," but one also speaks of an "atmosphere of violence" or of "holiness," and one even speaks of the "atmosphere of a boudoir," of a "petit-bourgeois atmosphere" or of the "atmosphere of the 1920s."

Building on these everyday experiences and phrases, the concept of atmosphere has by now been developed into a scientific concept.[3] The special feature of this concept, but also the theoretical difficulty, is that it refers to a typical intermediate phenomenon. Atmospheres are something between subject and object: they can be characterized as quasi-objective feelings which flow out indeterminately into space. Equally, however, they must be characterized as subjective, in that they are nothing without an experiencing subject. But it is precisely in this being-in-between that their great value lies. They link together what has traditionally been separated as the aesthetics of production and of reception. Of course, atmospheres can be produced, and there are highly developed arts the subject matter of which is specifically the production of atmospheres. This is done with the aid of entirely physical, technical means, which, however, do not act as causal factors producing effects, but as generators of atmospheres. The art of the stage set is paradigmatic of this type of access to atmospheres. On the other hand, atmospheres are experienced in terms of the affects they arouse, and one can only tell which type of character they have by exposing oneself to them in bodily presence, in order to feel them in one's own disposition. That is the classical aspect of reception aesthetics.

As means of studying atmospheres, contrastive and ingressive forms of experience have been proposed: the specificity of atmospheres is best experienced when their characteristics stand out – not when they have lapsed into something which surrounds us uniformly and inconspicuously. They are experienced, therefore, through contrast, when one is in atmospheres which cut across one's own mood, or upon entering them, through the switch from one atmosphere to another. Atmospheres are then experienced as "impressions" (*Anmutungen*),[4] that is, as a tendency to induce a particular mood in us.

With regard to production aesthetics, as I have mentioned, atmospheres are investigated in terms of what produces them: objects, the qualities of objects, arrangements, light, sound, etc. What is decisive, however, especially for thing-ontology, is that we are concerned here not with the properties a thing has, by which it is defined and distinguished from others, but rather with the qualities through which it radiates out into space. More precisely, what is important is to read properties as ekstases,[5] that is, ways in which a thing goes out of itself and modifies the sphere of its presence. The study of ekstases is especially important for the arts of design and stage scenery, since what matters there is not so much the objective properties and functions of things as their scenic value.

The aesthetics of atmospheres, which has its beginnings in ecological aesthetics,[6] has developed into a rehabilitation of the original approach of Alexander Gottlieb Baumgarten, that is, aesthetics as *Aisthetik*, as a general theory of perception.[7] It has subsequently proved its revelatory power in a series of case studies dealing with, for example, the atmosphere of a city, light as atmosphere, the atmospherics of twilight, the atmosphere of church spaces, music as atmosphere and, finally, studies of atmospheres of interpersonal communication.[8]

The aesthetic conquest of acoustic space

From the time of the ancient Greeks, music as artistic theory has been concerned with knowledge of notes and of what notes are, and was defined in terms of harmonic distances from a ground note or tonic, by intervals. This conception of music strikes us today as incomprehensibly limited. In contrast, in the twentieth century and beyond there has been an immense expansion of the musical material, unfolding in many dimensions. One may indeed speak of a conquest of acoustic space. The path has led from tonality via chromaticism to a gradual enlargement of the acoustic material which is admitted into music, right up to pure noise and din. If, to begin with, this material involved intermediate notes in the intervals of chromaticism, and the generation and inner life of sound itself – blowing, bowing, plucking – and if interest in the individuality of instruments, in their voices, grew, so also did the importance of sound. Then, with the distortion of instruments by banging and scratching their sounding bodies, and ever-new percussion instruments, an abundance of noises, not just notes, was admitted into music. Finally, with tape recorders, everyday noises, street scenes, natural sounds and the acoustic world of factories were incorporated in music. Today, the technique of sampling is making every kind of acoustic material available to compositions.

In addition to this enlargement of the musical material, a fundamental change or, more exactly, an enlargement of the essence of music, should be noted. Until well into the last century the dogma still prevailed that music is a temporal art. Its true nature was to be found in the time-figure, in a unity of the musical event which transcended the moment. From the basic demand for cadence and the return to the tonic, via melody and theme and the structure of movements up to the unity of the symphony, music was constituted by succession bound together as a unit. Even in the 12-tone music of a Schoenberg, through adaptation of fugal technique, the essential nature of music was seen to lie in the unity of the successive. This conception of music was not super-seded, but was undoubtedly relativized, by the discovery of music as a spatial art and its development more or less explicitly as such in modern music. That music fills spaces, and that, through resonance and reverberation, space is an essential element in its effect, has always been known. What was now dis-covered was that the individual note, the ensemble of notes, and also the sequence of notes or, better, the sequence of noises, have spatial structures, constitute figures and ensembles in space. That had not been previously a topic for music. Probably it was only the modern electronic techniques of the reproduction and production of music which made this area easier to manip-ulate, and thus attracted attention to it. The ability to make a sound buzz in space like an insect, and perhaps make it rise up over a muffled layer of sound and burst into sparks like a firework – these are possibilities which are only made available by technology and which direct our attention to something which, in a certain way, has always been a part of music. Even the Greek terms for high and low, namely ὀξύν and barún, the equivalent "pointed" and "heavy, spread-out," indicate this. In modern music, however, compo-sers began to work quite consciously on the spatial structure of music, partly by using conventional instruments and partly by electronic installations, enabling space to be acknowledged as a constituent dimension of musical organization. In some cases, this dimension can become the actual dimen-sion of the musical work in which, logically, anything like a beginning or an end, or a principle of a structure transcending time, can no longer be expected.

It is precisely music's tendency to become a spatial art which has brought it into the sphere of an aesthetics of atmospheres. The spaces involved here cannot be simply identified with geometrical space – or only, at most, with topological space. To be sure, in musical space there are directions, there is something resembling structures and also a kind of spacing-apart, but none of this is strictly in the sense of a separation, but rather in the form of changing, interpenetrating, emerging, and vanishing figures. Moreover, this space is experienced affectively: the spread-out as heavy and oppressive, the up-rising as lightening and joyous, the splitting-up as funny, and so on. Taking both these aspects together, one realizes that musical space is, to be exact, bodily space expanded, that is, a feeling-out into space which is shaped and articulated by music.

This discovery, that music is the fundamental atmospheric art, has solved for music theory an old, always intractable yet inescapable problem, the question of what the so-called emotional effect of music actually consists of. In contrast to the helpless theories of association, or the theories that have tried to see imagination as an intermediate link, the aesthetics of atmospheres can give a simple answer, that music as such is the modification of bodily felt space. Music shapes the feeling of the listener in space; it intervenes directly in his or her bodily economy. Practitioners have made use of this long before the theoretical insight was achieved: as early as the silent film, spatial and emotional depth was imparted to the image by music. The later film music followed this practice. In a radio play or feature one refers, in German, to an *atmo*, i.e. music or, more generally, acoustic events, inserted beneath the action in order to endow the spoken words with atmosphere. In a similar way, an atmosphere is generated in bars by a particular sound, and one's presence is made more agreeable at airports, in underground subways or at the dentist, or more cheerful and active in department stores or hotel foyers, by muzak.

What is true of atmospheres in general is daily reality in the case of acoustic atmospheres: the characteristics of a space are responsible for how one feels in that space. It has been discovered that the feeling for a homeland is mediated significantly by the "sound" of a region, and that the characteristic feeling of a lifestyle, of an urban or rural atmosphere, is very significantly determined by the related acoustic space. That is to say that what a landscape is can no longer be restricted to what one sees, and that town planning, for example, can no longer concern itself only with noise avoidance or noise protection, but must take account of the character of the acoustic atmosphere of squares, pedestrian precincts, and whole towns.

Music and soundscape, music of the soundscape

If the music of the last century extended the acoustic space by expanding the sound material to include technical sounds and everyday samples and even noise, and if music has finally evolved from a temporal art into an art which consciously shapes affective spaces, it has been met halfway in this conquest of space by a quite different development. I am referring to the World Soundscape Project founded by Murray Schafer in the 1970s.[9] In this project the world of natural sounds, the acoustic life of a town and the acoustic characteristics of technology and work were explored and documented, and finally compositions were produced from the material obtained. Acoustic and sound engineers collaborated with musicians or themselves became composers. What, viewed from the side of music, was a development leading to a widening of the musical material was, viewed from the side of the soundscape, a discovery of the musicality of the world itself. Of course, it has long been recognized that birds, and also whales, have their own music. But something more was involved here – the discovery of the acoustic characteristics or, more exactly, of the characteristic structures of living spaces, whether natural ones like the sea, the forest and other landscapes, or the living spaces of towns and villages. It became apparent that, even to document such acoustic worlds, compression and composition were needed, in order to communicate them to someone who was not at home in these regions. What could be more obvious than to make this compressing and composing into an explicit creative act, and through this both to come to meet music and to combine with it? John Cage's work, *Roaratorio*, is an example of the latter.

It is here that the work of Sam Auinger and his various partners, in particular Bruce Odland, is to be situated. More than any other, he enables us to participate through his productions in the great concert of the world. That, admittedly, is not easy, and for modern people, whose everyday listening is in reality a "listening-away," it is probably necessary to pass through music such as Auinger's in order to discover and appreciate this great concert. His procedure is different from Cage's, for example: it does not involve assembling files of sounds and noises and then composing from them by sampling. Rather, the transformation of given noise into music takes place on the spot, in actu. The noises are tuned using resonance bodies, usually a resonance pipe; that is to say that they are perceived in the form in which they cause the resonance body to vibrate with them, and are thus mediated by the natural frequencies of this body. This is an extremely interesting procedure. It reproduces in material form what might be regarded as the origin of music altogether: the transformation of noises into tones by tuning (that is, by the *signatura* of the resonance body). It is worth asking ourselves, in the light of the experiences afforded us by Auinger et al., whether our listening to the great concert of the world may perhaps consist in such a tuning of the noises which press in on us – a tuning performed by our ear itself. Is it not also the case that the visual sense constructs a relatively ordered spectrum of colors from the chaos of optical frequencies in the world?

Of course, tuning does not flatten out all noises into a series of tonics and harmonics. Rather, some of them – depending on their amplitude – retain their independent life. In this way Auinger et al. achieve what the collaborators on the Soundscape Project called the difference between tonality and characteristic event. Tonality is the basic mood of a landscape, a town, a harbor, and characteristic events are rare and distinctive noise-bundles which make up, so to speak, the physiognomy of a landscape. Such events may be – in Auinger's music as well – the sounding of a signal whistle or the screech of a train's brakes. What is also important is the continual appearance of the human voice, not in its linguistic articulation but as idiom, as the characteristic sound of a language. Auinger does not even shrink from occasionally using the sound of a classical musical instrument.

What is produced in this way might be called a piece of the great concert of the world of which Jakob Böhme speaks – admittedly not as God might hear it but as tuned for our ears, and therefore converted into music. And yet we, too, will hear the music out there in the world differently, in the way Auinger et al. make possible for visitors as listening participants in their installations: in Grand Central Station/New York, in the Haus der Kulturen in Berlin or, more generally, in an airport, on the motorway, in a pedestrian precinct. However, if one empathizes in this way, if one understands what is to be heard all around as music, that is, if one assimilates it through resonating in the manner described by Jakob Böhme, then, admittedly, one performs a redefinition of what music is: it is the play of acoustic events in a space stretched out by a tonality.

We should not see these two developments – that of music in the twentieth century and that of the Soundscape Project and the connection between them – in isolation from the development of technology. If the unfolding of music as a spatial art is practically inconceivable without electronic techniques of reproduction and production, so also is the exploration of acoustic landscapes without electronic techniques of recording and reproduction. But the development of acoustic technology in the twentieth century has also had an effect which is quite independent of the developments just mentioned, namely the omnipresence of music. Music, which in the previous centuries of the European tradition was something bound up with festivities and special occasions, has become a cheap, universal consumer commodity. Via radio and television, music is available at all times, and through the acoustic furnishing of public spaces our acoustic environments are, as a rule, already occupied by music or at least permeated by it. And where that is not the case, modern people bring their own acoustic worlds with them, first with the Walkman, now with the MP3 player.[10]

What are the consequences of this development? In view of the last point, one might speak of an acoustic pollution of our environment.[11] But on the other side it must be said that the acoustic consciousness of the average person has undergone a significant development. This is to say not only that musical needs and, within them, acoustic demands have increased considerably, but also that for the broad population hearing as such has developed into a dimension of life and a sphere of satisfaction. Of course, it must be said that the noise of the modern world and the occupation of public spaces by music has also led to an average practice of "listening-away." But even so, hearing has developed from an instrumental sense – I hear something – to a manner of participating in the life of the world. As for music, it must be said that through the developments mentioned its boundaries have become blurred. If it defined itself at the beginning of European musical history precisely by setting such boundaries, the constant widening of its field is tending to make every boundary uncertain. What Thierry de Duve[12] said with regard to the plastic arts – that after Duchamp the basic question of aesthetics "What is beautiful?" has turned into the question "What is art?" – also applies to music.

Acoustic atmospheres

Perhaps the reference to acoustic atmospheres has given a provisional answer to this question. Provisional here means an answer which defines what is characteristic of musical experience for our time. One must allow for the fact that at other times after us, and perhaps soon after us, different answers will have to be given. What is certain, however, is that the great period of Platonism in music has come to an end. Plato criticized people who tried to use their ears to find out what harmonic intervals were.[13] And even Adorno was able to say that the proper way to listen to a symphony was to read the score. We are worlds away from that today. It may be open to question whether modern music can be adequately written down at all. It seems that sensuousness in music has undergone a rehabilitation and that, contrary to the entire Platonic period, music can be apprehended only by hearing. Perhaps one must even say that the true subject of music is hearing itself. It has been said of modern art in general that it is reflexive, that it makes art itself – its social position, its anthropological meaning, its pure existence as a phenomenon – into its subject. In the visual arts this reflexivity had a clearly demonstrable purpose. Many of its works were no longer concerned with representing something, but with representing the experience of seeing itself. This may have started with Turner and the Impressionists, but became quite clear in artists such as Joseph Albers, Barnett Newman, and Mark Rothko. In music this development may not have manifested itself so clearly because it is, in a sense, much more natural to music. In contrast to the visual image, it has always been clear that music is object-less, that it represents nothing. Of course, there has been pictorial music, program music. But it cannot be denied that these were really wrong turnings, and that in taking this direction music was putting itself in the service of something else. Kant already said that music was the language of feelings. Of course, one might also interpret that statement, in keeping with the usual semiotic conception of language, to mean that music designates feelings, and therefore represents them. But that is not what Kant meant, for, with regard to spoken language, he distinguished precisely the tone in which something is said from the content mediated by signs; it was this tone which enabled others to participate directly in the feeling of the speaker.[14] In music, as he understood it, this way of communicating feelings became autonomous. Today we have reason to universalize this thought. According to it, the decisive feature of music is its ability to thematize acoustic atmospheres as such. This would provide us with a very different definition of music to the one we encounter in the Platonic tradition. In that tradition, music was defined fundamentally by restricting the sound material or, more exactly, by restricting the acoustic space by which musical notes were defined. We can say today that we are dealing with music whenever, in an acoustic event, we are concerned with the acoustic atmosphere as such, that is, with hearing as such and not with hearing something. This needs further explanation. But it can be said straight away that, understood in this way, music no longer needs to be man-made.

What does it mean to say that we are concerned with hearing as such and not with hearing something? In asking this question one discovers first of all the extent to which hearing is generally object-related. I hear a car driving past, I hear the clock strike 12, I hear someone talking, I hear a mosquito, I hear a ship's foghorn. This type of hearing is useful and plausible; it enables us to identify objects and locate them in space. But, in a sense, in this type of hearing, hearing itself goes unheard. Of course, instead of saying I hear a dog barking one can also say I hear the barking of a dog. But in reality that is a different type of hearing. No doubt, the barking belongs to the dog. It is one mode of its presence in space. But what is characteristic of voices, tones, sounds, is that they can be separated from their origins or can separate themselves from them, can fill space and move about in it practically like things themselves. To perceive acoustic phenomena in this way – and that means to perceive them as such, not as forms of expression of something – requires a change of attitude. We, the people of the twenty-first century, have rehearsed this extensively by using acoustic devices, in particular by listening with headphones. It is somewhat shaming for many of us that we only discover in this way that acoustic spaces are something autonomous, independent of things and not identical with real space. But, of course, acoustic space is also experienced in real space. Nevertheless, it is bodily space, the space of my own presence, which is constituted by the extent of my bodily awareness. In the hearing which does not jump over sound, voice, noise in order to reach the objects from which they may originate, the hearer experiences the voice, sound and noise as a modification of the space of his or her own presence. Those who hear in this way are dangerously open; they have let themselves out into a vastness where they may be assailed by acoustic events; bodily tunes may lead them astray, thunderclaps strike them down, whirring noises threaten them, a piercing tone injure them. To hear is to be outside oneself, and for that very reason may be the blissful experience of feeling that one is actually in the world.

These experiences must be undergone, they cannot be communicated verbally. However, there is a good analogy which may clarify what is meant. Descartes, in principle a philosopher who thought mechanistically, was once asked where someone who pokes a stone with a stick actually feels this stone. His answer, like that of the gestalt psychology of the twentieth century, was that one feels the stone where it is. This has been referred to as the making-corporeal of the stick, and that is not entirely wrong. But considered more precisely, what is involved is an expansion of bodily space. But how much more can it be said of hearing, than of poking with a stick, that in hearing we are outside. And our being-outside does not meet with voices, sounds, noises, but it is itself formed, moved, modeled, notched, cut, raised, pressed, expanded and confined by voices, sounds, noises.

The best model of hearing we have had up to now states that what is heard is re-enacted inwardly. This was the resonance model of hearing, and derived some plausibility from the well-known experience that, in some way, one inwardly "sings along" with a melody one hears. But this model suffers from the misplaced topology of an inside and an outside, and quickly runs up against its limit in the complexity and strangeness of what is heard. One will hardly be able to "sing along" inwardly with the noises of a machine shop, with its humming, shrilling, screeching, and whirring. Moreover, one does not hear all these noises inside but, of course, outside. That which is brought to resonate here, and that within which these voices, sounds, and noises take place, is bodily space itself. Admittedly, this experience occurs rarely or, more exactly, rarely in a pure form, for, in a sense, it is the basis of every experience of hearing. Normally, however, the "I" does not lose itself in hearing, but preserves itself by distancing the voices, sounds, and noises in their sources, and thus by jumping over the experience of what is in between.

Conclusion

In conclusion, we should return to the beginning. The development of music in the twentieth century has led to a situation in which music has itself become a constituent of the environment. In being functionalized as an aspect of interior design – one speaks of acoustic furnishing – it has to some extent been reduced to the atmospheric. On the other hand, avant-garde music on one side and the Soundscape Project on the other have elevated the acoustic atmosphere to the essence of music. In this way the voices of things and the concert of the world have attracted growing attention, and hearing has gained importance for life. If all this is taken together, then ecological aesthetics in the acoustic sphere becomes not merely a supplement to natural-scientific ecology, but rather has a task of its own, which is to know, preserve, and shape acoustic space. The question as to what a human environment is presents itself here as a question about the characteristics of acoustic atmospheres. Here, too, it is necessary to go beyond the merely scientific approach, which can comprehend noise at most in terms of decibels, and to ask which acoustic characteristics the spaces we live in ought to have.

Notes

1 Jakob Böhme, *Sämtliche Schriften. Faksimile-Neudruck der Ausg.*, Stuttgart, W.-E. Peuckert, 1955ff. [1730], vol. VI.
2 This impression evaporates on closer study. It emerges that examples from the field of design play a central role. Gernot Böhme, "Index over de aestetiske exempler i Kants 'Kritik der Urteilskraft,'" in *Kritik*, 105, 1993, 79–80 and *Kants Kritik der Urteilskraft in neuer Sicht*, Frankfurt/M., Suhrkamp, 1999.
3 Gernot Böhme, *Atmosphäre. Essays zur neuen Ästhetik*, Frankfurt/M., Suhrkamp, 1995. Michael Hauskeller, *Atmosphären erleben. Philosophische Untersuchungen zur Sinneswahrnehmung*, Berlin, Akademie Verlag, 1995.
4 Gernot Böhme, *Anmutungen. Über das Atmosphärische*, Ostfildern, edition tertium, 1998.
5 Regarding this term, see Böhme, *Atmosphäre*, Part III.
6 Gernot Böhme, *Für eine ökologische Naturästhetik*, Frankfurt/M., Suhrkamp, 1989.

7 Gernot Böhme, *Aisthetik. Vorlesungen über Ästhetik als allgemeine Wahrneh-mungslehre*, Munich, Fink, 2001.
8 The studies are in Gernot Böhme, *Architektur und Atmosphäre*, Munich, Wilhelm Fink Verlag, 2006.
9 As a more recent example of the extensive work of R. Murray Schafer, mention should be made of *Voices of Tyranny – Temples of Silence*, Indian River, Ont., Arcana Editions, 1993.
10 See the classic study on this subject: Shuhei Hosokawa, "Der Walkman-Effekt," in K. Brack et al. (eds.), *Aisthesis. Wahnnehmung heute oder Perspektiven einer anderen Ästhetik*, Leipzig, Reclam, 1990, pp. 229–51.
11 Critical of this: Hildegard Westerkamp, "Listening and Soundmaking: A Study of Music-as-Environment," in Dan Lander and Micah Lescier (eds.), *Sound by Artists*, Toronto, Art Metropole, 1990, pp. 227–34.
12 Thierry de Duve, *Kant after Duchamp*, Cambridge, MA, MIT Press, 1996.
13 Plato, *Politeia*, VII, 531a.
14 Immanuel Kant, *Kritik der Urteilskraft*, 3. Originalausgabe, 1799.

20 Seeing light

Is light invisible?

Despite our tendency to associate light with vision, the commonly accepted view is that light cannot be seen per se. We see things in light, but the light itself must strike something, even if it is only a small speck of dust, in order to be seen. This is a rather odd, indeed paradoxical hypothesis, given that light is defined specifically as that which we see. Viewed from the perspective of physics, light is electromagnetic radiation within the visible wavelength band. Thus light is related to vision by its very nature and is not, to be precise, a purely physical phenomenon, even from the viewpoint of physics. This relationship between light and vision is even closer at the phenomenological level. In his *Theory of Colors*, Goethe defines light, to the extent that he is willing to articulate definitions at all, as "nature acting according to its laws upon the sense of the eye."[1]

Strangely enough, we encounter the hypothesis that light is invisible in the writings of a physicist who is closely affiliated with anthroposophy. I am referring to Arthur Zajonc and his book titled *Die gemeinsame Geschichte von Licht und Bewusstsein*.[2] In order to prove that we cannot see light, he built a box into which he projected light from a projector and then, with the aid of some kind of device, captured the light inside the box in a volume, so that it did not come in contact with the walls. Oddly, Zajonc tells us nothing about the technology used to confine the light. I presume that it involved electromagnetic walls of the kind employed in confining plasma. Yet the effect achieved by this modern magician is all the more striking for that. When one looks through a hole in the side of the box, one sees absolutely nothing. Everything is dark. And one might well ask, "Of course, how could one see anything?" When no light can escape, there will be nothing to see!

The non-phenomenological character of Zajonc's approach consists in the fact that, quite apart from the matter of vision, he knows what light is, namely electromagnetic radiation within a specific frequency range. Physics offers good reasons for claiming that this radiation is inside the box and that we cannot see it. Yet from a phenomenological point of view, there is no light here at all if light truly is nature acting according to its laws upon the sense of the eye. As long as we are speaking of light as a phenomenon, it is absurd to say that one cannot see light itself. Evidently, Zajonc cannot deny his professional roots in physics, regardless of how concerned he may be with phenomenology.

Yet perhaps we should not be so quick to reject the view that light as such cannot be seen with reference to the definitions of light. It could well be that there is something missing in these definitions, something that is needed in order to enable light to appear at all. Let us call that something a *medium*. Goethe took a medium into account in his *Theory of Colors*, to the extent that he regarded light, darkness, and dimness (of dawn or dusk, for example) as necessary prerequisites for light. Dimness is a medium for manifestations of light, particularly for atmospheric ones. And Aristotle's theory of visual perception[3] also depends on a medium, namely the diaphanous, the translucent, without which nothing at all could be seen. Actually, neither of these theories can be used to refute the hypothesis of the invisibility of light. For no matter how closely Goethe focused on the theory of colors, he would never have contended *that seeing light means seeing colors* in every case. Aristotle's theory of perception is not concerned with seeing light but rather with seeing something. Still, we should take the reference to a medium seriously! Certain phenomena involved in the manifestation of light clearly suggest that we can see light only when it appears in a medium.

Light rays, luminous bodies

A classical view of the way we see light, and one that has played an important role in cultural history, is that of light as rays. The idea that we can see rays of light is responsible for the commonly accepted view that light radiates as beams. From the standpoint of physics, that is pure nonsense, of course. Yet it is a useful heuristic model for optics. Where does the idea of light rays originate? The answer must be that we see them. Quite apart from the fact that rays appear to radiate from light sources, especially small ones, in photographs taken at night, and quite apart from the same effect evoked by stars that appear as more or less as dots of light with extended points, the most obvious examples are the bands of light that pass through church windows and the powerful rays that break through towering cumulus clouds. Within the context of neo-Platonism, both of these manifestations of light have often been interpreted as signs of God's attention to the world.

Both of these phenomena demonstrate that light requires a medium in which it can be seen if it is to be visible. This is particularly evident in the bands of light seen in churches. Because this are so close, we see that what makes light visible in this case are particles of dust floating in the air. When struck by light, the particles become luminous. The same effect can be observed in the mist that rises from a waterfall. These beams of light from the sky demonstrate their reliance on matter – perhaps for the very reason that they are distant from us – in that they can be interrupted. The reason for this, of course, is that the light that passes through spaces between the clouds strikes mist or wisps of cloud at some points but not at others. Thus we see light in the form of rays, yet this phenomenon itself clearly shows that what we see is really the glowing matter that is struck by light as it moves along its way.

These relationships can be demonstrated even more dramatically by projecting light in a relatively dust-free room. The light may be visible immediately in front of the projector and on the screen, but not in between the two. If one holds one's hand or a piece of paper between the projector and the screen, it will catch the light, showing that it is present between the two points as well. Ergo, light itself is not visible.

But wait! In the case of the bands of light in churches and the rays we see through openings in the cloud cover, our observation of the dust particles or cloud formations breaks down the appearance of the rays and thus does not merely *explain* it. This is obviously an either–or phenomenon. Either one sees the whole, i.e. rays, or one sees the detail, particles of dust and wisps of cloud. One might decide therefore to define light rays simply as one type of light phenomenon. And that would mean that we can indeed see light as such, namely as light rays.

On the other hand, one might also decide to recognize seeing luminous bodies as a genuine way of seeing light. To return to the example above, we do not actually see the luminous dust particles as dust particles; we simply know that that is what they are. We see points of light. And the same goes for luminous wisps of cloud or bands of mist. Of course we can also see them as something substantive, as a cloud or a patch of fog – but only when they stand out against a dark background. Otherwise, their material character disappears behind the manifestations of light, which they are. Thus these luminous bodies are light manifestations of a unique kind. Although they are flat surfaces or bodies, because they are expanded and limited, their only imposing feature is the fact that they are luminous. Viewed as a phenomenon, they are nothing but light manifested in a certain way. Their material character is not evident.

Taken together, both of these phenomena, luminous bodies and light rays, embody the primal image of the manifestation of light as it is deeply and indelibly imprinted in our culture: the sun – and the stars as its smaller sisters. And that brings us to the most dramatic way in which we see light as such – by looking at a source of light.

Sources of light

In our world of consumer goods, we generally refer to artificial sources of light as lamps. It is a maxim of the lighting industry that such light sources should be designed to prevent blinding glare to the extent possible. Lamps should be seen and viewed as luminous objects. The unique aspect of these objects is that they show themselves. Thus they differ from all other objects we see only in light but not as light. This difference shows once again that luminous bodies must be addressed as genuine evidence of the fact that we see light. The principle of avoiding blinding glare is intended to ensure that this self-manifestation is relatively discreet. For it is true that these luminous bodies relate to us, that they approach us, and this means that they are fundamentally aggressive. Absence of blinding glare is meant to allow the subject to look actively at the light source, which also means being able to see it as an object.

The situation is different with light sources we still experience as natural light sources, including in particular the sun. I deliberately chose the word experience, for looking at the sun is definitely not the same as *seeing the sun*. Yet this kind of seeing is in a very real sense a seeing of light. It is important to express oneself carefully here. Being blinded is an experience of light in which we are overpowered by light, which means that we have no possibility of objectifying what we see. Blinded vision is a kind of seeing, but not of seeing something.

This brings us to a crucial point from which the paradoxical view that we cannot see light can be approached. The direct confrontation with light, as we experience it paradigmatically when we look at a blinding source of light, is not seeing something but seeing per se, a kind of unintentional vision. It is in this sense that we use the word *see* when we say of someone who has had eye surgery that he can see. We do not mean that he can distinguish objects – that may often be possible only after gradual adaptation and training. What we mean is that he can perceive light through his eyes. This example shows that the simple process of seeing light is much more fundamental than that of seeing something. And the contention that we cannot see light results from the fact that we ordinarily think of seeing as seeing something. But light is not a something, and the perception of light is primarily unintentional vision.

Recognition of this principle of unintentional vision opens up a broad field of light phenomena for consideration.

Pure light phenomena

Expressions such as "I see a light" and "I see light" are quite common. They denote a non-material manifestation of light – ordinarily against a dark background. The stars and, under certain circumstances, the moon belong to this category. Yet we know all too well that they are objects, and in the case of the moon, we can usually see signs of its physical form in its shaded areas. This can be quite different in the case of a glow-worm, when it is encountered unexpectedly. Actually, pure light phenomena also include such things as the *aurora borealis*. The diffuse character of this manifestation, its shapelessness, the fact that it is manifestation of light without a source makes it a prototype of the manifestation of light per se. The events that herald and follow the day are similar in nature. Before sunrise and after sunset we occasionally see pure light in the sky. But pure light phenomena exist not only in the sky but in the world of objects as well. While light normally allows us to see things, it can also liberate itself from them as well. Although it appears on things, it gains a certain degree of independence which allows it to dance around them. Hence the shine and shimmer of things. And because of certain material conditions, which do not enter into the phenomenon itself – such as surface properties that create interference or the atmosphere that surrounds objects – it may happen that an aura of light envelopes these bodies – a halo. Incidentally, *halo*, in its meteorological sense, refers to the wreath of light we often see around the sun or the moon, which is the result of the refraction and reflection of its light by ice crystals in the atmosphere.

A characteristic feature of all of these manifestations of light is that light appears in them as such. Thus they have become prototypes of *manifestations* in general in the course of cultural history of manifestations, that is, in which something of a spiritual nature becomes perceptible through the senses without becoming a being or thing. God can appear as pure light, or we see an individual's holiness as a halo or a person's character as an aura.

None of the light manifestations mentioned up to this point are a totality. As diffuse as they may be, they are limited in space. The artist James Turrell created pure light manifestations as totalities. I call them totalities because their uniformity encompasses the entire field of vision. And that is why they cannot be depicted in pictures. Turrell erected installations at the Sprengel-Museum in Hanover in which visitors stuck their heads into a kind of large hair-drying hood that enclosed the head from the front rather than the rear. Inside the hood, the viewer saw nothing but a certain color, not on a surface but as a spatial phenomenon by which he felt surrounded. It was intermittent light, and its intensity and rhythm could be regulated with dials placed in front of the visitor. And the visitor had to regulate them if he was to maintain a sense of being in control of the situation. For what he saw, an indefinite vast color space without contours of any kind, was capable of cutting him off from reality completely. As fascinating as the vision was, one felt threatened by it. Every sun-worshipper has experienced a related phenomenon. When one closes one's eyes in the face of bright sunlight, one finds oneself in a hemisphere of intense reddish-orange to yellow hues without a visible horizon.

This phenomenon of light vision in which one sees nothing but light and experiences oneself as present in a light space leads us finally to the fundamental experience of light as lightness.

Lightness

The word *light* tempts us to regard the phenomenon we are talking about as a something, an entity. We have long since learned from physics that this entity cannot be comprehended as separate from the way in which we cause it to appear. Yet the complementary forms of existence in which it is conceived – wave or corpuscle – have not contributed to refutation of the concept of light as substance. A wave is surely not a thing, but it is a quantity of energy which, as widely distributed as it may be in a given space, can still be imagined as compact. This idea of light as a something is reinforced by the fact that we generally assume that light originates from a source. We have already demonstrated that light is not a something and does not necessarily always originate from a source, and that it is often irrelevant to light phenomena. The phenomena in which objects play no role and light itself cannot be objectified are actually fundamental to vision in general. We see only in the presence of light – that is, when it is light. Therefore, I have proposed that we speak of lightness rather than light.[4]

Goethe frequently used the terms light and lightness synonymously. The analogy is light is to dark as lightness is to darkness. Yet he speaks of lightness as an entity *"das Helle"* and not as a quality *"die Helle,"* a distinction he also makes in the famous passage on the primal phenomenon of color in his *Theory of Colors*:

> One such primal phenomenon is the one we have discussed above. We see on the one side light, lightness and on the other darkness, the dark. We place dimness in between the two, and from these opposites, with the aid of thoughtful mediation, colors develop, also in opposition.[5]

With *lightness*, Goethe refers to an objective representation of light, such as a piece of white paper. But I use the term *lightness* to indicate that it does not have the character of a thing but of a freely floating quality or, as Hermann Schmitz has expressed it, a quality that flows indeterminably into the vastness. A quality of what, one might ask? And to that I would answer: of space.

We are in good company with this concept of light as lightness. Aristotle regarded light (φως) as a state of excitation, an *Enaergeia* of the translucent. Although he conceived of the translucent (το διαφανης) in concrete terms as something like water or air, its material character is not the point. The crucial insight is that Aristotle regarded light as a state of excitation in space which is shared by the perceived object and the perceiving subject. And thus Aristotle aptly described the fundamental experience of seeing. We see things in light, which means when it is light.

The crucial factor is that the recognition of lightness is the fundamental experience of seeing. Not only the man who had had eye surgery cited in an example above, but every one of us sees that it is light before recognizing things. We wake up late, for example, one morning, and what we notice first, before we make out objects and symbols, is that it is light. The fundamental experience of morning is the recognition of beginning lightness – and this progress toward light begins before the sun appears and we have reason to attribute the lightness to the presence of the sun. And Aristotle speaks in the same terms, incidentally. Not that the sun is the cause of lightness, through its radiation, for instance, but that lightness is the presence of the sun, its παρουσια.

The fundamental manifestation of light is lightness. We see light primarily as lightness, and that means first and above all that we see things in lightness.

In this sense, lightness is a transcendental manifestation, namely a manifestation that makes the appearance of other manifestations possible. Lightness as a manifestation is a totality, like the light manifestation in the installation by Turrell cited above. We see lightness not as an object but as a quality of the space that surrounds us. Lightness can even be restricted in a spatial sense, in that it is bordered by darkness. But this presupposes its totality. Lightness is also limited in a temporal sense. As a state, it can begin and end. We refer to this temporally limited lightness as *day*. Thus the day is also a light phenomenon. That is very vividly expressed in the Bible:

> And God said, Let there be light: and there was light. And God saw the light, that it was good: and God divided the light from the darkness. And God called the light Day, and the darkness he called Night. And the evening and the morning were the first day.

> (Genesis, 1:3–5)

Remarkably, God did not create the stars, the sun and the moon until the fourth day. They are ordering factors and signs signifying the passage of time, not sources of light.

Because we see light as lightness, it is a fundamental fact of perception. But what do we really see when we perceive lightness. Is it primarily the fact that it is light? We have already established that lightness is not an object. But does the perception of lightness have content nonetheless?

We must begin by saying that lightness, as a quality, appears in varying degrees – it can be more or less light. Brightness, as a degree of lightness, varies between absolute darkness and an excessive degree of brightness that cannot be defined precisely. Excessive lightness is brightness of a degree that is unbearable for human beings. It is interesting to note that there is a certain asymmetry between darkness and brightness. Absolute darkness may also be unbearable under certain circumstances, depressing and frightening. But it can also just as easily have a calming, comforting effect. But, above all, darkness does not intensify infinitely but reaches an absolute zero point. That may be the reason why we do not regard darkness as a quality in its own right but as a deficient mode of lightness, as the absence of light.

Lightness and darkness can be distributed unequally in space or, to be more precise, the degrees of lightness may vary in time and space. This produces an abundance of additional light phenomena, such as dawn and dusk, shadows, pale light, and twilight. Yet these phenomena reveal another fundamental aspect of the perception of lightness, meaning that they emerge from it in many different ways through specific articulations. To perceive lightness is to perceive space. Light as lightness creates space.[6]

This statement must be expressed more precisely. We do, of course, experience our existence as being in space even in the dark. But, in the dark, that space is partly close and confining and partly endlessly vast, devoid of orientation, and all-consuming. The reason for these fears of confinement and vastness, to which sick people are particularly vulnerable, becomes clear when it grows light. Lightness creates space as *spatium*, as the space of distances. There is closeness and distance even in the space of lightness, but they are *definite* closeness and distance, and for that very reason space is created as free space, namely as distance from things and room for action.

The fact that lightness creates space is largely responsible for the emotional – or, to cite Goethe, the sensual–moral effects of lightness. As things become visible in the light, they also appear to us in space. Space is not created by the distances between things. It is merely articulated by things, as are the degrees of light and dark. Things appear to us as closer or more distant in the infinite vastness of illuminated space. We can wander within this space with our eyes, and that – even more than the concrete possibility of moving, which also exists in the darkness – gives us the feeling of being in space.

Moreover, lightness, which we have defined as a quality of space, appears to be devoid of qualities itself. In a certain sense, it is the neutral medium of visibility and, as such, is secondary to the visible. This may be the reason why Goethe hardly gave any consideration to light at all. In his view, light was a precondition of color phenomena that could not be analyzed further. Colors are the *deeds and sufferings of light*. They appear when light is forced into a corner, so to speak. And that is also why Goethe avoided speaking specifically of colored light. We also recognize that lightness enables us to see things, including their colors. Yet we must concede that the experience of lightness itself exists in colorful variations. The installations by Turrell discussed above call our attention to this in a particularly striking way. In everyday life, however, we register such experiences as special kinds of *illumination*. This expression emphasizes above all the difference between daylight and artificial light. Color-neutral daylight is compared to artificial light, which is capable of changing colors. We use similar expressions when we see colorful light in a natural setting, at sunrise or sunset, for example. We then speak of special lighting effects and mean that light causes things to appear in different colors than those to which we are accustomed and also that this effect is not a totality but relates only to certain parts of our surroundings.

It is by no means easy to find order in these relationships. The fact that daylight seems colorless to us may be attributable to its ubiquity and its unremarkable ordinariness. It would be something like the smell of one's own home that one no longer notices. But this would mean that the natural colors of things are merely their conventional colors: the colors they display in daylight. That would be in keeping with the view that we regard moonlight as a special kind of illumination, namely a whitish, silvery light. This underscores the conventional character of our concept of the natural colors of things in combination with the colorlessness of daylight. But his convention has its roots in our own human nature. We are by nature diurnal beings after all.[7]

Thus as the lightness of day, daylight has a normative function. Compared to daylight, all other forms of lightness, i.e. colored types of light that make things appear to us in different coloration, are regarded as mere illuminations.

Illuminations

Illuminations are color-modified lightness, and they involve above all additive color mixtures. In other words, the color of the illuminating light is added to the intrinsic color of the illuminated object. This presupposes, of course, either that the illuminated objects emit light themselves or that the components of light needed for the intrinsic color of the object, the components that allow this intrinsic color to appear, are present, either in the basic lightness of daylight or in the illuminating light itself. To express this with a negative example, if we illuminate a blue surface in the dark with orange light, the blue color will not appear, of course.[8] If, however, the conditions mentioned above are met, the result will be an additive, intensity-enhancing color mixture composed of the intrinsic color and the illuminating color.

This effect also plays a role in normal daylight in the form of interaction between neighboring colors, as each of these colors emits the light of its color, meaning that they illuminate each other. This effect, known as the Bezold spreading effect, was used by the Impressionists, who created mixed colors that appear suspended in space with their pointillist style of placing small dabs of pure color close together. Josef Albers was particularly interested in these relationships of interaction between intrinsic colors.[9] He also made use of another effect which Goethe advanced as one of the primary elements of his theory of colors – simultaneous contrast. When looking at a colored surface for an extended period of time, the eye supplements the color with its complementary color in its surroundings. This effect is in dynamic opposition to the Bezold spreading effect, and the two can offset one another in a certain phase.

The fact that illumination places the intrinsic colors of objects in a different light or, in other words, can make them appear in a different color, plays an important role in marketing, merchandise presentation, and interior architectural design. The desired colors must always be seen as dependent upon the color of the illumination. This can be used to achieve positive effects, but people will always want to eliminate its negative consequences as well. Thus savvy customers occasionally walk outside a shop in order to see what color a particular article of clothing *really* is.

But that actually plays a role only when new purchases are made. As a rule, we *know* what colors things are. And that means, in concrete terms, that changes in the colors of individual objects are largely eliminated in everyday life. We ordinarily do not notice them at all, no more than we notice the effect of color shadows that plays such an important role in Goethe's theory of colors. As Goethe clearly knew and expressed by using the term *didactic* in the title of a section of his *Theory of Colors*, one must first learn to see color shadows. The fact that we ordinarily see color shadows as gray or black and do not notice these changes in the colors of objects is not simply a matter of inattentiveness. It is a perceptual effect in its own right. Gestalt psychologists speak of color constancy. Here as well, our perception is not merely passive but actually normative and functional. Thus corrections are made during the process of perception, corrections that seek conformity with simple, familiar forms, constancy in things and thus constancy in color as well.

For this reason, we ordinarily do not see the coloration of an illumination on an individual object – although that would certainly be possible. Instead, we see the coloration of the illumination as a tinting of the whole. What is involved here is a totality effect. The whole world or an entire scene is given a certain tint by the illumination.

This tinting of the field of vision, which is laid like a veil over an entire landscape or scene and amounts in essence to a change in the color composition of the whole, is the true field of the *sensual–moral effect of color.*[10] The color of an individual object may very well influence an individual's mood. The sight of a yellow dress might cheer up an observer, for example. But the sensual–moral effect is actually something more atmospheric. It is the emotional tint of the space in which one finds oneself that determines how one feels. This view is supported by the fact that Goethe repeatedly used examples of an overall, or total impression in his discussion of the sensual–moral effects of colors. In his remarks on blue, for example, he writes: "To a certain degree, rooms papered entirely in blue appear wide, but actually empty and cold."[11]

He could not have demonstrated this effect very well with reference to a solitary cup. In order to demonstrate the sensual–moral effects of colors, he refers in practically every case to colored glasses through which the world can be viewed. I quote here from the passage on yellow:

> This warming effect [of the color yellow, G.B.] can be experienced most vividly by looking at a landscape through yellow glass, especially on grey winter days. The eye is pleased, the heart swells, we feel a sense of good cheer. We seemed to be touched by immediate warmth.[12]

Incidentally, producers of tour buses also take advantage of this effect. By tinting the glass in bus windows, they make the world look friendly and cheerful.

This tinting of the world is different than the effect achieved by illumination. A tinted plate of glass acts as a filter, meaning that the change in color is the result of subtraction. Colored illumination produces a change in color through addition. Yet the effect is the same in either case. Through colored illumination or through the effect of looking through colored glass, all of what is seen takes on a tint that turns the diversity of what is seen into a unified whole. We are familiar with such unity in our surroundings primarily through daily, seasonal, and weather-induced changes in illumination. Thus we speak, for example, of a scene in the evening light, of the light of a landscape swept by a warm föhn wind, of the light of sunset, of dusk, or dawn. Each of these forms of illumination imbues the landscape with a specific, characteristic atmosphere.

And thus we have arrived at another fundamental way in which light is seen. We see light as colorful illumination in the characteristic modification of the total color impression. Evidently, we are sensitive not only to colors and differences between colors but also to color modifications as well.[13] And the totality of these modifications is obviously sensed as a visible emotional quality. Therefore, it is appropriate to say that illuminations are perceived as atmospheres.

This fact has immense practical significance for the illumination of rooms, squares, and cities. With the aid of illumination, entire scenes can be overlaid with a color-modifying hue, lending a characteristic mood to the whole. The paradigm for this technology is provided in this case, as it is true of the creation of atmospheres in general,[14] by the art of the stage designer. Illumination of one of the most important resources he has for creating a certain climate or, as I would say, a certain atmosphere on the stage.

Dealing with light

Thus at the end of our discussion of how we see light we arrive at the question of how we deal with light. Indeed, this question has now become the primary motive for posing the question of how we perceive light. The range of possibilities offered by lighting technology has expanded tremendously and continues to develop so rapidly that the question of light is no longer, as it was in Goethe's time, a question of how we accommodate ourselves to nature by educating ourselves. It has now become a question of how we deal in practical terms with light and its effects. The fundamental fact of perception with which I introduced the subject of *seeing light*, namely the recognition that *it is light*, has turned for most people into *I make it light*. And the concept of illumination at which I ultimately arrived no longer has the character of a question about a fact of nature – *Look at the beautiful light there in the West!* It has become instead the question of a practical approach to the use of light. We illuminate individual objects in technical arrangements, in merchandising aesthetics or in scenes, for example, in the theater or in the staging of city sights. Lighting design has become a discipline in its own right, with applications in architecture, urban planning, interior architecture, theater and museums, marketing and advertising, including even the presentation of goods at the point of sale. But the art of lighting also plays an important role in photography, film, and television, of course. Practically nothing of what we see in our daily environment – by which I mean capitalist, technical civilization – is seen simply as it would appear of its own accord, i.e. in the plain light of day. In other words, everything is staged with the aid of lighting technology. Everything is illuminated with aesthetic intent. This form of staging is devoted in the spirit of generalized merchandise aesthetic to the creation of an atmosphere, a basic emotional mood, in which we perceive the world, the specific scene and the things in it, especially the goods. In order to understand this practical use of light and criticize it, if need be, it is important to consider what it means to see light.

Notes

1 Translated from J.-W. Goethe, *Zur Farbenlehre.* Werke FA 23,1, Frankfurt/M., Suhrkamp, 1991, p. 25
2 Arthur Zajonc, *Die gemeinsame Geschichte von Licht und Bewusstsein*, Hamburg, Rowohl, 1994.
3 His theory of perception is discussed in the essay on the soul. Specifically, in 18b of the Bekker edition.
4 In my essay entitled "Licht als Atmosphäre," in R. Olschanski (ed.), *Licht-Ortungen. Zur Reflexion des Sichtbaren*, Bodenheim, Philo, 1988, pp. 32–43.
5 Didactic Part, paragraph 175.
6 See the detailed discussion with references to applications in architecture and design in my essay titled "Licht und Raum. Zur Phänomenologie des Lichts," in R. Betrus and R.-M. Peplow (eds.), *Symbolisches Flanieren. Kulturphilosophische Streifzüge*, Hanover, Wehrhahn, 2001, pp. 142–57. Also in *Logos 7*, 2001/2002, 448–63. An abridged version is printed in *Deutsche Bauzeitung*, 2002, 3, pp. 43–5.
7 This statement, like Paracelsus' contention that we are beings of air, requires some modification. Physiologists specialized in the senses have pointed out that there is both day vision (aided by cones) and night vision (aided by rods). I wish to thank Marion Bernhardt for calling this to my attention.
8 These relationships are investigated and presented in greater detail in the Colour-LightLab Project, www.farblichtlabor.ch; last accessed 20 May, 2016.
9 Josef Albers, *Interaction of Color. Grundlegung einer Didaktik des Sehens*, Cologne, DuMont, 1997 (originally published by Yale University Press, 1963). Josef Albers, who was a good didactic thinker, must be read as a theorist with caution. He frequently refers to color effects, which are true visual phenomena, as optical illusions.
10 Translated from J.-W. Goethe, *Farbenlehre, Didaktischer Teil*, 6th sec., FA 23/1, p. 247ff.
11 Paragraph 783.
12 Paragraph 769. See Paragraph 784 for blue, paragraph 798 for purple. It has been established, by the way, that Goethe and those among his contemporaries who were interested in an aesthetic lifestyle observed landscapes through a Claude glass, a glass with a brownish tint, in order to obtain a view of the kind rendered by the landscape painter Claude Lorrain. In our age, sunglasses presumably produce a similar effect.
13 Goethe's quoted remark on the effect of the yellow glass should actually be corrected at this point. One cannot overlay a landscape with yellow color through a yellow glass. What actually occurs is that the blue and near-blue components of the perceived colors are filtered out. And in a winter landscape, this would leave large areas of yellow.
14 See my book titled *Aisthetik. Vorlesungen über Ästhetik als allgemeine Wahrnehmungslehre*, Munich, Fink, 2001.

21 The phenomenology of light

What is light as a phenomenon rather than considered as a physical fact? I can give an answer to this question by recalling a very fine definition of color that Goethe gave us. He said that color was the laws of nature in relation to the sense of the eye.[1] Perhaps the phenomenology of light consists exactly of that – studying those selfsame laws of nature relative to the sense of the eye. In which case, you soon discover that you cannot stop at colors. You have to take all the phenomena of light into account – the glow, the brilliance, the flickering, shadow, and lots of other things besides. Allow me to cite an example to highlight what's special about studying nature this way relative to the sense of the eye.

We are indebted to Arthur Zajonc for his recent endeavor to bring together in a single book all the experiences and concepts we connect with light, from physics to mythology. The full title of the book is *Catching the Light: The Entwined History of Light and Mind*,[2] admittedly a strange, wholly unphenomenological phrase. He is of the view that we can't see light. To prove it, he constructed a box with light projected into it from a projector but with a device inside locking up the light within a volume so that it can't touch the walls. It is alarming that Zajonc does not tell us anything about the technique he uses for the locking up. I assume it involves walls of the electromagnetic kind used to lock up plasma. But the effect that this modern magician generates is all the more impressive. When you attempt to look into the box through a hole in the side, you see nothing at all. It's completely dark. Of course it is, you say, how could you see? If light can't get out, you won't see anything, either.

The un-phenomenological aspect of Zajonc's procedure is that he already knows independently of sight what light is, i.e. electromagnetic radiation within a certain frequency spectrum. For sound physical reasons one can assert of this radiation that it is in the box and we can't see it. But, phenomenologically, there is no light here if light is the laws of nature relative to the sense of the eye. When you talk about light as a phenomenon, it is absurd to assert that you don't see light itself. Zajonc evidently cannot disown his professional background in physics, much as he dallies with the phenomenology as well. The assertion that light itself cannot be seen does, however, have a more phenomenological (if weaker) meaning. It is a familiar experience that, when you enter a dark church, you see the rays of natural light as such because of the particles of dust they encounter. It is a gross abstraction to assert that, if the church interior were completely clean, there would be nothing but pure blackness between the church window and the floor that the light strikes. It is certainly a false conclusion that we cannot see light as such except when it strikes bodies, since when we look directly into a light source, we do see, though without seeing anything particular. We realize that, even in this weaker assertion that we cannot see the light itself, there lies a presupposition or, more accurately, a bias or presumption, that makes it pure tautology. It is assumed in fact that seeing always means seeing *something*, or, to be more precise, something tangible, an object. The assertion is then nugatory, since light is not something tangible.

This brings us to the fundamental phenomenological fact relative to light. Light as a phenomenon is primarily and actually brightness (i.e. bathed in light).[3] We shall see that there is a wealth of other typical light phenomena besides, but the brightness is fundamental. When I open my eyes on a day I've overslept, the first thing I notice is that it is already light. Noticing brightness is primary and fundamental. It precedes every individual perception of (for example) color, shapes, and things. I perceive all these things, but in the brightness. And this becoming aware of brightness is the basic experience of light.

Before going on to individual phenomena to do with light, I should like to say a little about the fundamental nature of light *qua* brightness. Philosophers are inclined to describe this nature as *transcendental*. "Transcendental" means "a precondition of the possibility of." If we are tackling phenomenology of visual phenomena, we notice that light *qua* brightness is attributed a particular role among these visual phenomena. We only see everything we see as long as it is light (i.e. in brightness). Brightness is thus a precondition of the possibility of any seeing. It is transcendental for seeing. This special position of brightness in the visual field contains great potential that can be exploited artistically. One might say that appearance itself becomes, as it were, manifest in brightness, or conversely – that brightness co-features in every visual appearance. The converse is that seeing is always a topic of light art. That light *qua* brightness is a precondition of the possibility of seeing at all was formulated by Plato in his famous metaphor of the sun, where he says:

> If in the eyes the faculty of seeing exists and one wants to make use of it, and if also there are colors (in the things) – if then there is a third entity missing which in particular is fit for it, then – you know it well – the sight will not see anything and the colors will be unseen.
> What is it you are talking about?
> That, what you call light, was my answer.[4]

It is clear that Plato is referring to light *qua* brightness here. Brightness is itself a phenomenon, but a phenomenon with transcendental meaning. Brightness is what turns sight into a real capability in the first place, and enables visible things to be seen in reality.

We shall need to refine this fundamental insight below, but equally need to hold tightly to it. That is because light is not the only precondition of visibility. Darkness is another. True, light and darkness are asymmetrical. Light is a precondition for seeing at all, whereas darkness (interacting with light) is a precondition for our seeing *something*, i.e. that there are things such as definition, articulation, and determinateness.

Cleared space

The first effect of light *qua* brightness is that it opens up a space. In a sense, that space is even created by light. To grasp that, it has to be explicitly understood what we mean by space here. It is not of course mathematical or physical space, for example, which can, where necessary, be measured in the dark as well. It really involves space as we experience it, and even then only in the sense of a particular experience. We also know of purely acoustic spaces, for example from the experience of space from listening through earphones. In the nature of things, such spaces have nothing to do with light. The space that light creates is the space of distances, extent, remoteness from me. It is best described as "clear (i.e. illuminated) space." Its most distinctive feature is its illuminant quality ("making bright"). You get spatial experiences in darkness as well, with a space seeming close and oppressive, or conversely you can lose your bearings in the indefinite extent of darkness. The typical transformation that takes place with brightness is that you find yourself topographically situated (you have measurable bearings) and at the same time the space around you allows room for free movement. For this reason, it seems appropriate to call this type of space, thinking of the concept of a clearing, "clear space." Like a clearing in a wood, created by cutting down growth, it is determined by distances between things (boundaries) and on the other hand the possibility of free movement. A characteristic feature of the space created by brightness is that the possibility of moving around it is not just an opportunity for actual movement, but also an opportunity for possible movement, i.e. moving merely with the eyes. You can rove around a clear space with your eyes. That such movement is indeed a significant spatial experience is evident from the fact that a corresponding experience cannot be had from photos. The reason for this is that the focuses in a photograph are fixed at the moment of photographing, whereas in a clear space where I find myself, I can look around in the depths, i.e. not just let my eyes rove from one object to the next but also fix my attention at changing distances. It could be that this possibility of roving around the depth of the space with the eyes is crucial for our feeling of finding ourselves in a space.

The primary emotional experience of a clear space is one of security and freedom. It is of course possible for threats to come at you in a clear space, but the fundamental experience is nonetheless that everything is at a distance and that this distance means security and freedom of movement. This factor of security in a clear space acquires the quality of a safe harbor if the clear space itself is demarcated, i.e. is clearly distinguishable for the indeterminate space of darkness. Conversely, seen from the clear space, the darkness becomes an area of indeterminate threat. Clear space without limits is daylight. We see here that daylight also has a spatial character. Daylight as such is unlimited, but it has nonetheless to initially unfold in the morning, and in the evening withdraw and disappear.

One might ask whether the experience of clear space demands that a source of light be part of the experience. As we have linked the phenomenon of clear space to the simple experience of brightness and in particular the possibility of roving around the space with our eyes, we probably have to say that the perception of a light source does *not* belong to this experience. That is a very important conclusion, because it is all too easy to see light in physicality terms as an emanation from a light source. Even the Greeks, who were not yet into thinking physicalistically – whereby I mean Aristotle specifically – always related light to its source. Aristotle defines light as *parusia*, i.e. the presence of the sun or fire, as appropriate. However, we can (or, given our experiences with indirect light and *light-like objects*, need to) make a distinction. By light-like objects, I mean things like luminous ceilings, or even stained-glass windows in medieval churches, which people frequently say are experienced as luminous walls. It may be disputed whether we ought to talk of an experience of brightness without a source of light here. The decisive thing is that we can notice the brightness as such, which is why the experience of clear space is not dependent on the perception of a light source, either. Only an effect would give one cause for thought. There is a kind of shadowless illumination of a space – for example, as generated by computer lights – whereby space loses depth and becomes more or less flat. In some circumstances, space can acquire a distinctly surreal quality because you can no longer assess the relative remoteness of things (i.e. indirectly assess the depth of the space properly) just by looking. That would mean, however, that what I previously described as freedom in clear space (the possibility of roving around spatial depth with the eyes) could be something to do with the definition provided by shadow. That would mean that an indirect ancillary experience of the light source, i.e. through shadow, is important for the full experience of clear space.

Light space

Light creates space. That was our first conclusion. And I called the space that brightness opens up clear space. Clear space is a space where I happen to be. I experience my presence in the space in a certain fashion by means of the brightness. Now, it is also possible, however, to see a space created by light virtually from outside, more or less as an object. This phenomenon has really only come to the fore as a result of light engineering and been demonstrated to us by light art in fascinating and often disconcerting ways. I am thinking here first and foremost of James Turrell's work. Turrell has created various forms of installation in which light spaces – for example, blocks of stone or pyramids – seem to float in darkness. Tellingly, preparation is always necessary by passing through a light trap and entering a dark room. At the Museum of Modern Art in Frankfurt, for example, after a while you become aware of a picture – or perhaps more accurately a block, because the object appears to have a certain depth – floating on the wall made of even, colored light. If you move closer, the room opens in the wall into a hazily illuminated *exterior*, of indeterminate depth. Experiencing these light spaces has a dreamlike quality, probably because it is so wholly independent of any experience of objects. Possibly that is what makes such experiences so inherently disconcerting, and for many people even disturbing. The impression of three-dimensionality you get from these constructs depends incidentally also on how close you are to them. If you saw them from a greater distance, they might perhaps simply look like light sources. This phenomenon, which is demonstrated by art in pure form, could play a part in other light experiences in impure form, i.e. mixed and with other phenomena superimposed on it. That is how an illuminated stage area and, shall we say, an office space you look into from a dark street, are perceived in this ambivalence of light space and light source. The magical element inherent in this experience may be linked to the fact that you experience these light spaces as potentially clear spaces you might step into, as it were, thereby being to that extent displaced from the outside where you find yourself into an imagined inside. Obviously that is what illuminated shop windows depend on, at least as long as the difference in light between the shop windows and the street remains intact. Walter Benjamin was probably thinking of this in saying in his *Arcades Project* that the goods were presented as on a stage.

Holograms also belong to the genre of light spaces. These create figures formed of light floating freely in space by manipulating interference phenomena. Remarkably, this effect has so far scarcely been applied in advertising, though it has in entertainment. At Disney World, for example, you can see ghosts made of light sitting at a table. Such phenomena may most clearly demonstrate that light is a transcendental phenomenon, a manifestation that conjures up the Other but also manifests itself. This self-manifestation of light, we may now say, can at the same time also simulate a something – a block of stone or a pyramid, as in Turrell, or robber figures in Disney. That is why these phenomena are only manifestations – which is also how specters are described – i.e. manifestations without a something that becomes manifest.

Generally, *mere manifestations* still have to appear on something real, at least a projection surface or monitor or some such. However, there is no doubt with either *laterna magicas* or the virtual world of monitors that the constructs we see there are actually made of light, i.e. "photo-graphs," to take the Greek etymology literally. The more we can forget the piece of reality they become manifest on, the more fascinating they become.

Lights in space

The prototype for the phenomenon of "lights in space" is the night sky. Here, too, you could say more or less that light opens up the space, but it is a different light and a different space we need to bear in mind than was the case with brightness. The stars are lights that happen to be in dark space, but they remove from dark space the oppressiveness and uncertainty wherein you might lose your way. However, they do not turn that space into one where you can judge distance, such as happened with brightness. The stars do not permit an estimation of distances, but they do give space a shape, articulating it by setting directions. That also gives a degree of security for spatial feeling, and we do know that navigation is possible precisely due to that kind of security. The space articulate by stars nonetheless remains dark. That also means that lights in space are not actually perceived as a source of light, even though – it may be observed in passing – they supply the clearest proof that you can see light. To be perceived as a light source, they would have to shine on something. In actuality they *are*, of course, sources of light and in toto they do brighten the night a little. This brightening is more or less imperceptible, and is not readily perceived as originating with the stars (it is quite different with the moon). We should define the phenomenon of *lights in space* in terms of this very quality. They are perceived as points of light rather than sources of light, even though that is what they really are.

This type of light is moreover not always pure when found in nature. Blurred sight or mist is, after all, enough to give stars a corona. The phenomenon of *lights in space* is very well illustrated by the phenomenon of glow-worms. In this case, there is the additional effect of movement and particularly an irregularly hovering movement. That makes clear what you sense in looking at the stars – you experience lights in space as something autonomous, with a life of its own. That may be an effect of "bodily communication"[5] (to use Hermann Schmitz's term) or even identification – certainly, at any rate, an inclination to transpose oneself to the location of the light in space and look down on our world from there.

Light and things

I have discussed three main phenomena of light – clear space, light space, and lights in space – without going into the relationship between light and things. That seemed necessary to me because the traditional bias is to associate light closely with bodies. This bias is closely related to another, to the effect that light as such cannot be seen. Yet for phenomenology, it needs stating explicitly that light is also manifest as such. The manifestation of light on bodies, important though it may be in practice, is in contrast only secondary – as an indirect becoming-manifest.

Notes

1 Johann Wolfgang von Goethe, "Farbe ist die gesetzmäßige Natur in Bezug auf den Sinn des Auges," *Farbenlehre*, HA XIII, p. 324.

2 Arthur Zajonc, *Catching the Light. The Entwined History of Light and Mind*, Oxford, Oxford University Press, 1993.

3 Gernot Böhme, "Licht als Atmosphäre," in Gernot Böhme, *Anmutungen. Über das Atmosphärische*, Ostfildern, Tertium, 1998.

4 Plato, *Republic*, 507d/e, my translation [G. B.].

5 Hermann Schmitz analyzed the importance of bodily communication for perception in *System der Philosophie*, vol. III, 5, Die Wahrnehmung §242, Bonn, Bouvier, 1978.

Bibliography

Atmosphere as a fundamental concept of a new aesthetics

(English) *Thesis Eleven*, 36, 1993, 113–126. [published with permission from SAGE Publications Ltd]
(German) in *Atmosphäre. Essays zur Neuen Ästhetik*, Frankfurt/M, Suhrkamp, 2009 [1995].

Atmosphere as an aesthetic concept

(German and English) "Atmosphäre als Begriff der Ästhetik," *Daidalos*, 68, 1998, 112–115.
(German) also in *Architekturzentrum Wien (Hg.), Sturm der Ruhe. What Is Architecture?* Salzburg, Anton Pustet, 2001, pp. 37–41.

The art of the stage set as a paradigm for an aesthetics of atmospheres

(German) "Die Kunst des Bühnenbildes als Paradigma einer Ästhetik der Atmosphären," in Ralf Bohn and Heiner Wilharm (Hg.), *Inszenierung und Vertrauen. Grenzgänge der Szenographie*, Bielefeld, Transcript, 2011, pp. 109–117.
(English) "The art of the stage set as a paradigm for an aesthetics of atmospheres," *Ambiances* [online], 10 February 2013, http://ambiances.revues.org/315

Kant's aesthetics: a new perspective

(English) *Thesis Eleven*, 43, 1995, 100–119. [published with permission from SAGE Publications Ltd]
(German) in *Kants Kritik der Urteilskraft in neuer Sicht*, Frankfurt/M, Suhrkamp, 1999.

On beauty

(English) in *3deluxe*, Amsterdam, Frame Publishers, 2008, pp. 174–184.
(German) "Schönheit–jenseits der Dinge," in N. Adamowsky u.a. (Hersg.) *Affektive Dinge. Objektberührungen in Wissenschaft und Kunst*, Göttingen, Wallstein, 2011, pp. 198–212.

On synesthesia

(German and English) "Über Synästhesien," *Daidalos*, 41, 1991, 26–36.

Contribution to the critique of the aesthetic economy

(English) *Thesis Eleven*, 73, 2003, 71–82. [published with permission from SAGE Publications Ltd].

Aesthetic knowledge of nature

(English) *Issues in contemporary culture and aesthetics*, 5, 1997, 37–45.
(German) "Eine ästhetische Theorie der Natur," in G. Böhme, *Natürlich Natur. Über Natur im Zeitalter ihrer technischen Reproduzierbarkeit*, Frankfurt/M, Suhrkamp, 1997 [1992].

Nature in the age of its technical reproducibility

(English) *Issues in contemporary culture and aesthetics*, 1, Maastricht, Jan van Eyck Akademie, 1995.
(German) "Natur im Zeitalter ihrer technischen Reproduzierbarkeit," in G. Böhme, *Natürlich Natur. Über Natur im Zeitalter ihrer technischen Reproduzierbarkeit*, Frankfurt/M, Suhrkamp, 1997 [1992].

Body, nature, and art

(German and English) "Der Leib, die Natur und die Kunst," in Ludwig Forum für-internationaleKunst (Hrsg.), *Natural Reality. Künstlerische Positionen zwischen Natur und Kultur*, Stuttgart, DACO-Verl., 1999.

Nature as a subject

(English) in catalog *Making nature*, NikolajUdstillingsbygning, Nikolaj Copenhagen Contemporary Art Center, 2002, 8–14.

The atmosphere of a city

(English) *Issues in contemporary culture and aesthetics*, 7, Maastricht, Jan van Eyck Academie, 1998, 5–13.
(German) "Die Atmosphäre einer Stadt," in G. Böhme, *Architektur und Atmosphäre*, München, Fink Verlag, 2006.

Atmosphere as the subject matter of architecture

(English and German) "Atmosphere as the subject matter of architecture"/"Atmosphäre als Gegenstand der Architektur," in Ph. Ursprung (Hrsg.), *Herzog & de Meuron: Natural History/Naturgeschichte*, Montreal, Canadian Center of Architecture and Lars Müller Publishers, 2002, pp. 410–417.

Staged materiality

(German and English) "Inszenierte Materialität," *Daidalos*, 56, 1995, 36–43.

Architecture: a visual art? On the relationship between modern architecture and photography

(German and English) "Architektur: eine visuelle Kunst?," in Ralf Beil and Sonja Feßel (Hg.), and Andreas Gursky, *Architektur*, Ostfildern, Hatje Cantz, 2008, pp. 24–31.

Metaphors in architecture – a metaphor?

(English) in Andri Gerber and Brent Patterson (eds.), *Metaphors in Architecture and Urbanism. An Introduction.* Bielefeld, Transcript, 2013, pp. 47–57.

Acoustic atmospheres

(English) "A Contribution to the Study of Ecological Aesthetics," *Soundscape*, 1(1), 2000, 14–18.
(German) "Akustische Atmosphären," *Hochschule Musik und Theater Zürich*, Akustik- Symposion, Zürich, 21.8.2006, pp. 3–8.

Music and architecture

(English) "Atmospheres: The Connection between Music and Architecture beyond Physics," *Metamorph Focus. 9.* International Architecture Exhibition, Venice, Fondazione La Biennale, 2004, Bd. 1, pp. 110–114.
(German) "Atmosphären: die Beziehung von Musik und Architektur jenseits physikalischer Vorstellungen," in Chr. Metzger (Hrsg.), *Musik und Architektur*, Saarbrücken, Pfau-Verlag, 2003, pp. 98–108.

The great concert of the world

(German and English) "Das große Konzert der Welt," in Carsten Seiffarth and Martin Sturm (Hrsg.), *OK offenes Kulturhaus Oberösterreich, A hearing perspective*, Wien, Folio Verlag, 2007, pp. 3–15 (German); pp. 47–58 (English).

Seeing light

(German and English) "Licht sehen," in Ulrich Bachmann (Hrsg.), *Farben zwischen Licht und Dunkelheit*, Sulgen/Zürich, Verlag Niggli, 2006, pp. 115–135.

The phenomenology of light

(German and English) in Ursula Sinnreich and Kulturbetriebe Unna (Hrsg.), *James Turrell, Geometrie des Lichts*, Ostfildern, Hatje Cantz, 2009, pp. 69–78.

Index

Printed in the United States
by Baker & Taylor Publisher Services